Exploring Music

The Science and Technology of Tones and Tunes

The Royal Institution lecture theatre
immediately following one of the lectures.

Exploring Music

The Science and Technology of Tones and Tunes

Charles Taylor

Former Professor of Experimental Physics
The Royal Institution

and

Emeritus Professor of Physics
University of Wales

Institute of Physics Publishing
Bristol and Philadelphia

British Library Cataloguing in Publication Data

A catalogue record for this book is available from the British Library

ISBN 0-7503-0213-5

Library of Congress Cataloging-in-Publication Data are available

Published by IOP Publishing Ltd,
a company wholly owned by the Institute of Physics, London

IOP Publishing Ltd,
Techno House, Redcliffe Way, Bristol BS1 6NX, UK

US Editorial Office: IOP Publishing Inc.,
The Public Ledger Building, Suite 1035, Independence Square, Philadelphia, PA 19106

Typeset in TEX at IOP Publishing Ltd
Printed in Great Britain by J W Arrowsmith Ltd, Bristol

Contents

Foreword ix

Introduction 1

1 What is Music? 4

1.1 Introduction 4
1.2 The nature of sound 4
1.3 Sound waves in solids and liquids 8
1.4 What makes a sound musical? 9
1.5 Pitch and frequency 10
1.6 Detecting ultrasonic waves 12
1.7 Pure tones 13
1.8 Music from wooden blocks 15
1.9 The first family of musical instruments 15
1.10 The ear and hearing 16
1.11 Measurements on hearing 19
1.12 More questions about the nature of music 24
1.13 Music and information 27
1.14 What can be varied? 28
1.15 Harmony and discord 30
1.16 Beats and difference tones 34
1.17 Psycho-acoustic complications 37
1.18 More about the part played by the brain 39
1.19 Conclusions 41

2 The Essence of an Instrument 43

2.1 Introduction 43
2.2 Starting a note 43
2.3 Natural frequencies 45
2.4 Keeping a note going 46
2.5 Making the sound loud enough: instruments of the first family 47
2.6 Making the sound loud enough: instruments of the second family 52

2.7 Other consequences of using sound boxes 54
2.8 Vibrations of air in a tube 56
2.9 Edge tones 58
2.10 Harmonics: instruments of the third family 61
2.11 Reeds 63
2.12 Analysing musical sounds 65
2.13 Why do reeds produce so many harmonics? 67
2.14 How we perceive mixtures of harmonics 68
2.15 Harmonics of strings 72
2.16 Keeping string vibrations going 75
2.17 A contemporary mechanical instrument 79
2.18 How notes change with time 80
2.19 The all-important beginning of a note 83
2.20 More about the origin of transients 86
2.21 Conclusion 87

3 Science, Strings and Symphonies 89

3.1 Introduction 89
3.2 Patterns of vibration of plates 91
3.3 Patterns of vibration of air in hollow bodies 95
3.4 The bodies of stringed instruments 99
3.5 Bowed instruments 103
3.6 Making a violin 105
3.7 Can science help? 109
3.8 Testing in the concert hall 111
3.9 The wolf tone 114
3.10 The Catgut Acoustical Society 115
3.11 Hand-plucked strings 119
3.12 Keyboard-operated plucked strings 121
3.13 Keyboard-operated struck strings 124
3.14 The pianoforte 124
3.15 Piano touch 127
3.16 Conclusion 128

4 Technology, Trumpets and Tunes 131

4.1 Introduction 131
4.2 What happens at the end of a tube? 132
4.3 Vibrations in tubes open at both ends 133
4.4 Vibrations in tubes closed at one end 135
4.5 Privileged frequencies 136
4.6 Vibrations in conical tubes 138
4.7 Harmonic recipes in the woodwinds 140
4.8 Edge-tone instruments 143
4.9 Wind-cap instruments 145

4.10 Mouth–reed instruments 145
4.11 The functions of side holes 147
4.12 How the sound gets out 147
4.13 Keeping the vibrations going 149
4.14 The functions of keys 154
4.15 Another view of vibrations in tubes 156
4.16 Transition instruments 158
4.17 Turning a tube into a trumpet 160
4.18 Valves and slides 162
4.19 Harmonic recipes in the brass family 165
4.20 Organ pipes 165
4.21 The mechanism of an organ 170
4.22 The voice 173
4.23 Conclusion 178

5 Scales, Synthesisers and Samplers 179

5.1 Introduction 179
5.2 The purpose of scales 180
5.3 Equal tempered scales 182
5.4 Consequences of temperament 185
5.5 Electronic synthesis 186
5.6 Analogue synthesis 188
5.7 Sampling 195
5.8 Digital techniques 197
5.9 Computer synthesis 199
5.10 Digital synthesizers 200
5.11 The concept of MIDI 203
5.12 Why synthesise anyway? 205
5.13 Mechanical instruments and their successors 207
5.14 Conclusion 209

6 Reflections, Reverberation, and Recitals 211

6.1 Introduction 211
6.2 Everybody must be somewhere 212
6.3 Loudness versus intelligibility 214
6.4 The work of W C Sabine 216
6.5 What time of reverberation is desirable? 218
6.6 Placing the absorbent 220
6.7 Placing the reflectors 221
6.8 Some unfortunate consequences of reflection 221
6.9 Some subjective problems 225
6.10 Methods of acoustic design 227
6.11 Adjustment of the acoustics 230
6.12 Symphony Hall, Birmingham 232

6.13 Noise in buildings 234
6.14 Conclusion 236

Appendices

A Holographic Interferometry 237
A.1 Introduction 237

B Pitch and Frequency 241
B.1 Introduction 241
B.2 Systems of pitch notation 241
B.3 Ratios for the Just diatonic scale 242
B.4 The use of cents in frequency measurement 243

Acknowledgments 245

Bibliography and Suggestions for Further Reading 249

Index 251

Foreword

Every year, ever since 1826, the Royal Institution has invited an eminent scientist to deliver a course of lectures at Christmastide in a style 'adapted to a juvenile auditory', to use the words of Michael Faraday, who initiated the tradition. In practice this means that the lecturer will be confronted with an enthusiastic and critical audience ranging in age from under ten to over eighty and, in respect of scientific knowledge, from the relatively untutored child to staid professors of science, venerable Fellows of the Royal Society and a few Nobel Laureates, all of whom will expect the lecturer to say something that will interest them.

The present book is an expanded version of what Professor Taylor said and demonstrated during the highly successful series he gave over the Christmas season 1989–90. Charles Taylor is exceptional among scientists in that he is capable—as he puts it modestly—'of extracting some form of music' from some twenty instruments. And, apart from being a world renowned crystal and optical physicist, he has unique experience in reaching the hearts and stimulating the thought of young people. He has probably presented more lecture-demonstrations to children than anyone else in Europe. Regularly at the Royal Institution for the past twenty years he has addressed approximately one thousand (of the thirty thousand) children that throng every year to its theatre to be inspired, educated and entertained in ways of which Faraday would have approved. In his days (1965–83) as Professor of Physics at the University of Wales, Cardiff, he addressed children and adults all over the Principality and beyond. It is estimated that the aggregate viewing audience for his two series of Christmas Lectures broadcast by the BBC is in excess of 20 million! His live audience figure reaches well over 70,000.

Readers of this book will enjoy the pleasure and the understanding that Professor Taylor imparts to young and old alike when he explores, in his unique way, the magical quality and science of music.

Sir John Meurig Thomas, F.R.S.
Fullerian Professor of Chemistry at the Royal Institution
and Deputy Pro-Chancellor of the University of Wales.

Introduction

It is well known that many scientists, especially physicists, often have a great love of music, and some are very good instrumentalists. Yet, at school, science and music are usually taught as quite separate subjects. I was very fortunate to be taught both physics and music by the same teacher and so I began, at a very early age, to be interested in the connections. Indeed at one stage I thought that they were, in fact, part of the same discipline. So my preoccupation with the relationship can be traced back to that teacher—the late Clifford Early—who started off my interest some 60 years ago. In later years I found that most schools did not have such a civilised attitude and tended to keep music in a separate category from science. I decided therefore that, as soon as possible, I would try to remedy this defect.

So when, about forty years ago, it was suggested that I might give lectures to schools to try to encourage students to take up science, the topic of science and music seemed to be an obvious choice. Not only are the relationships fairly easy to indicate, but the joint topic lends itself very well to lecture demonstration.

In more recent times it became clear that lecturing to fifth and sixth forms is really too late and that it would be more productive to talk to the 7–11 age group, where decisions about future topics to be studied had not already been made. In this age group there is the added advantage that most children love making music and it is possible to use this in order to persuade them that science is fun and to persuade the teachers that you can interest children in science without necessarily being an expert.

Naturally I was delighted to find that, when the National Curriculum in Science was published, there was a whole section devoted to Science and Music.

In 1971 I was invited to give the Royal Institution Christmas Lectures on the topic of Science and Music. They were televised on BBC 2 and, subsequently, I wrote a book based on the lectures, published by the BBC, and given the same title as the lecture series—*Sounds of Music.*

When, in 1989, I was invited to give another series of televised Christmas Lectures on a similar topic I was astonished to find what enormous changes there had been in the field in the intervening 18 years. Nevertheless many of the ideas and demonstrations from the first series were still valid and, with the advantage of new technologies, could still be usefully included.

This book is a record of the second set of lectures, again using the same title as that of the series and following the same sequence as was followed in the lectures. Of course, in a book it is possible to expand some ideas beyond what was possible in the lectures themselves where strict time limits had to be imposed. One or two new developments that have occurred since the lectures were given are also included. And there is one topic that had to be omitted completely from the lectures—the influence of the acoustics of rooms on instrumentalists and on listeners to music. The final chapter of this book is devoted to that.

Science has often been described as the process of asking questions about the world around us. Doing an experiment is one of the most powerful ways of trying to find an answer. So our exploration of music in a scientific way will consist mainly of asking questions and then doing experiments to try to find answers.

One of the many problems of converting the substance of a series of lectures into a book is that of describing the demonstrations in sufficiently graphic detail that the reader can gain the same insights as an audience witnessing the live demonstrations. Fortunately some of the experiments are relatively easy to perform and so practical advice on performing them is included at various points.

Many people have helped in many different ways, both in the preparation of the lectures and of the book. I have tried to acknowledge them all in a separate section, but, inevitably, in a book of this kind I have drawn on the experience of many years and it has not always been possible to remember the source of an idea or demonstration. I hope therefore that any sins of omission will be forgiven and that the unrecognised originators will accept my gratitude.

There are however a few people who must receive special thanks. Lord Porter, P.P.R.S., was the person who invited me to give the first series of Christmas Lectures and later to become Professor of Experimental Physics at the Royal Institution, so enabling and encouraging me to use the incomparable resources of the Royal Institution on many subsequent occasions. His successor as Director of the Royal Institution, Professor Sir John Meurig Thomas, F.R.S., was the person who invited me to give the second series of Christmas Lectures on which this book is primarily based. He also made available space and equipment at the Royal Institution to enable me to prepare the lectures and has continued to support and encourage my lecturing ventures. I owe them both an enormous debt of gratitude.

I should also like to express my special gratitude to three mentors who have had a great influence on my career at three different stages. First the school teacher, already mentioned at the beginning of this note, who gave me such firm foundations in both physics and music. Secondly the late Professor Henry Lipson, F.R.S., who taught me both about research and the importance of communication, guided my first 17 years of University teaching, and was a source of help and encouragement until his death in 1991. Thirdly the late Dr C W L ('Bill') Bevan who, as Principal of University College, Cardiff not only supported my efforts in my main line of research (x-ray and optical diffraction), but also encouraged the development of the physics–music research and teaching group at Cardiff. He also supported to the hilt my involvement in the popularisation of science through public and schools lectures at a time when such activities were not as fashionable as they are today.

I also want to say a very special word of thanks to my wife who has supported me with unfailing love and encouragement, and has tolerated rooms full of books and apparatus in our house during the preparations for the lectures, and ever since, with extraordinary patience.

1

What is Music?

1.1 INTRODUCTION

It is common in science to find that we cannot find a direct answer to the primary question and have to start by asking a great many smaller questions, or, perhaps, by accepting simplistic answers to parts of the question. This is particularly true of the scientific aspects of music. There is no simple answer to the question 'What is music?' either from the scientific or the artistic point of view and so we must start with a great many peripheral questions.

Music is obviously a kind of sound and a sound is something that you hear. So it will be helpful to start with some ideas about sound, how it travels and how we hear it.

1.2 THE NATURE OF SOUND

That air is necessary for the transmission of sound was first demonstrated by Robert Boyle. His experiment was revived by one of the earliest scientists to make use of lecture demonstrations, Francis Hauksbee the elder, in 1705. John Tyndall regularly performed this classic demonstration in his lectures on sound in the Royal Institution round about the middle of the 19th century. The apparatus that he used, which was presented to the Royal Institution by Warren De La Rue, is illustrated in figure 1.1. A clockwork bell is hung on rubber bands inside a glass bell jar that can be evacuated. There are many ways of performing the experiment but my favourite is to start with the apparatus (the actual one used by Tyndall is still in existence and was used in the lectures) outside the theatre. An assistant pumps out the air and closes the tap, sets the bell ringing and then carries it in. The audience can see that the clapper is striking the bell but cannot hear it. The lecturer opens the tap to let in air and the audience can then hear the bell.

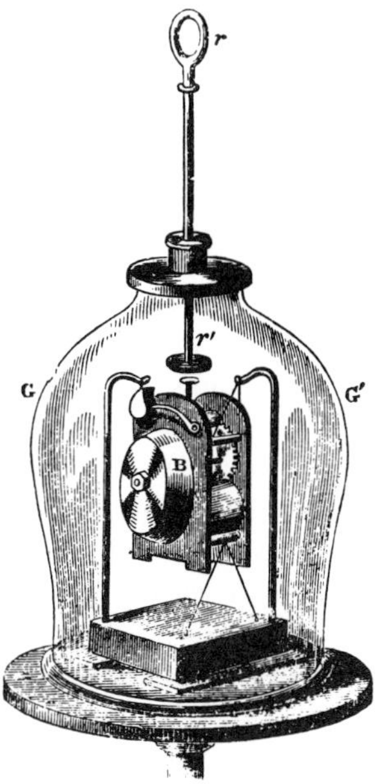

Figure 1.1 Apparatus for demonstrating that sound will not travel through a vacuum.

Thus we have demonstrated that air is necessary for the production of a sound. But what happens to the air when a sound is made? If a party balloon is inflated and then gently squeezed and released, the pressure (or 'squash') of the air inside can obviously be increased and decreased by a small amount. But no sound is produced. However if the pressure is changed violently by means of a pin, then there is obviously a loud sound. Further investigation shows that it is not the size of the pressure change that matters, but its speed. Indeed ordinary conversation involves changes of pressure in the air made by the voice of one or two parts in a million. And that is about the same as the change in pressure inside the balloon if a finger is gently pressed about 1/100 mm into a balloon that has been blown up to about 20 cm in diameter.

The fact that the pressure changes in speech are very rapid can be shown by connecting a microphone to a cathode ray oscilloscope. The oscilloscope is one of the most important aids in lectures on sound and music because one can see exactly what kind of pressure changes are involved in any particular sound. It can be introduced to a non-scientific audience as a device for drawing graphs.

If the scanning circuit of an oscilloscope is switched off to give a small spot of light on the screen and then a microphone is connected, the spot moves up and down in response to the pressure changes. When the scan is switched on to make the spot move across the screen at successive steps of increased speed, the graph of the pressure variation against time is drawn. The scale of the horizontal time axis depends on the speed of the trace (see, for example, figure 2.32 in Chapter 2).

Both the microphone and our ears are really pressure gauges. In both there is a thin diaphragm that responds to the changes: in the ear it is a natural membrane called the eardrum; in the microphone it is usually of metal. In the ear the mechanism of the middle and inner ear—which we shall return to later in the book—converts the pressure changes into nerve signals to the brain; in the microphone one of several electronic techniques converts the pressure changes into a varying electric current which can be fed to the oscilloscope and/or to an amplifier and loudspeaker.

But how do the pressure changes travel from the source of sound to the ear or to the microphone? Most people know the answer...they travel as waves. But what exactly is a wave? There are, of course, countless varieties of waves—sea waves, sound waves, light waves, seismic waves and so on. What do they all have in common? The key feature is that although something moves along, it is not the medium itself. We are all familiar with the fact that if a toy boat is lost in the middle of a pond, throwing a stone into the water near to the boat will create waves that spread out in circles from the point where the stone enters the water and, though the boat will bob up and down as the waves pass under it, it will not move very far from its original position. A good example is the famous domino experiment. A series of dominoes is set up in a line so that as any one falls over it knocks the next one down. A 'push' applied to the first will be transmitted to the last and could carry sufficient energy to turn on a switch, but the dominoes will have hardly moved from their original positions after the wave has passed. The two kinds of waves we have described are significantly different. The waves on water involve oscillations, or wobbles, about a fixed point. (Every point on the water surface moves up and down in a regular way as the wave passes.) The dominoes undergo a single displacement in turn. But, in each, energy may be transmitted over long distances without a wholesale movement of the material over that distance. I sometimes illustrate the same idea using a line of children and the hypothetical experiment of asking what will happen if I give the first one in the row a big push. John Tyndall used to do this in his lectures on sound at the Royal Institution over 150 years ago. Here is a short extract from his description in his book *Sound* (see figure 1.2).

> 'I have here five young assistants, A, B, C, D, and E, placed in a row, one behind the other, each boy's hands resting against the back of the boy in front of him. E is now foremost and A finishes the row behind. I suddenly push A, A pushes B and regains his upright position; B pushes C; C pushes D; D pushes

> E; each boy, after transmission of the push, becoming himself erect. E, having nobody in front, is thrown forward. Had he been standing on the edge of a precipice, he would have fallen over; had he stood in contact with a window he would have broken the glass;...'

Figure 1.2 'Boy wave' from John Tyndall's *Sound*.

Sound waves involve the transfer of pressure changes from point to point in the air and normally occur in three dimensions; i.e., a change of pressure at a particular point will normally spread out in all directions at the same speed.

In normal circumstances sound waves will travel in air at about 340 metres per second (about 760 miles per hour). The exact speed depends on the temperature and humidity and other factors

The front edge of the wave will be a sphere which expands in diameter over a period of time. The waves can be reflected, refracted and distorted in various ways. Some of these possibilities will be discussed in Chapter 6 under the general heading of the acoustics of rooms.

A demonstration of the way in which waves spread out from a point was performed during the lectures using a large section of the audience. The Royal Institution lecture theatre is semi-circular in plan and steeply raked.

> The children in one sector were asked to place their hands on their heads (to constitute light coloured objects for a television camera looking vertically downwards from the roof). They were then asked to lean forward and then to sit back in their seats a row at a time starting from the front. The roof camera view was recorded and played back to the audience and gave a realistic model of how a sound wave spreads out.

1.3 SOUND WAVES IN SOLIDS AND LIQUIDS

But sound does not travel only through the air—it also travels through solids and liquids. If you stand at one end of a very long wooden or metal fence and hit it with something hard, friends at the other end can hear the sound very easily, especially if their heads are placed in contact with the fence. And if you are swimming under water you can still hear noises. Sir Charles Wheatstone (the inventor of the concertina and populariser of the 'bridge' for making electrical measurements) performed a famous experiment to demonstrate how musical sounds can be transmitted through wooden rods. It was called 'Wheatstone's Telegraphic Concert'. Instruments in the basement of the Polytechnic (the site of the experiment) were connected by wooden rods to harps in a first floor hall. Listeners in the hall heard the sounds of the instruments in the basement apparently emerging from the harps, but those on the ground floor (through which the rods passed) heard nothing.

The version most often used at the Royal Institution is that due to John Tyndall. He made a hole in the floor of the lecture theatre and, in keeping with the traditions of the Royal Institution that nothing useful is ever thrown away, the hole remains to this day. A wooden rod passes down through the hole and the lower end rests on the sound board of a piano. When the instrument is played the sound waves pass up the rod and can be conveyed to the body of a harp resting on the upper end of the rod.

It is worth remembering that, when this experiment was first done, there were no gramophones or loudspeakers and the public were not familiar with sounds emerging from inanimate objects. John Tyndall remarked 'An uneducated person might well believe that witchcraft or "spiritualism" is concerned in the production of this music.'

Tyndall's description of the result of the experiment is a classic example of Victorian eloquence:-

> 'What a curious transference of action is here presented to the mind! At the command of the musician's will the fingers strike the keys; the hammers strike the strings, by which the rude mechanical shock is converted into tremors. The vibrations are communicated to the sound board of the piano. Upon that board rests the end of a deal rod, thinned off to a sharp edge to make it fit more easily between the wires. Through the edge, and afterwards along the rod are poured with unfailing precision the entangled pulsations produced by the shocks of those ten agile fingers. To the sound board of the harp before you the rod faithfully delivers up the vibrations of which it is the vehicle. This second sound board transfers the motion to the air, carving it and chasing it into forms so transcendently complicated that

> confusion alone could be anticipated from the shock and jostle of the sonorous waves. But the marvellous human ear accepts every feature of the motion, and all the strife and struggle and confusion melt finally into music upon the brain.'

I have included this quotation not just because it is such a splendidly evocative description of the experiment, but because it reminds us at this early stage in the book of the enormously important contribution to the hearing process that is made by the brain. So important is it that I shall tend to refer to the 'ear–brain system' rather than to the ear alone.

1.4 WHAT MAKES A SOUND MUSICAL?

Just as *beauty* is said to be in the eye of the beholder, so *music* could be said to be in the ear of the listener. A superb performance of, say, a Debussy prelude in a practice room may be considered an awful noise by a student in the next room who is trying to study the score of a Beethoven sonata. Similarly the hideous wailing of an ambulance siren might be music to the ears of those anxiously awaiting its arrival.

But let us start with a very simple but obviously totally non-musical sound. Physicists call the kind of sound that is produced by steam escaping from a valve, or by a television set that is not tuned to a station, 'white noise'. The oscillograph trace of such a sound looks like figure 1.3. The pressure is changing very rapidly, but in a completely irregular way. It is difficult to see any pattern in the trace at all. You could imagine the sound going on for ever, never changing and it is probably the most boring sound imaginable. But if its loudness is made to vary a bit and you use some imagination it begins to sound like the waves of the sea breaking on the shore. And it only needs the sound of a seagull to produce quite an effect on the members of an audience! Even if there is only a brief seagull sound to set the scene and then just white noise varying in loudness, they begin to think of lazy summer days sunbathing on the beach.

We have not changed the actual sound very much, but we have produced a profound effect on the listeners! So perhaps influencing listeners is one of the features of music?

In the lectures we set up four microphones of the kind used by sports commentators and asked four members of the audience to say the same words simultaneously. Each microphone produced a trace on a four-channel oscilloscope so that the audience could see all four traces simultaneously. The trace for speech looks superficially like that of white noise and each trace was different from the others. This immediately seems paradoxical! The fact that the traces look rather like white noise would suggest that no information is being conveyed and yet we know that this cannot be so as the traces are graphs of the pressure changes involved in speech.

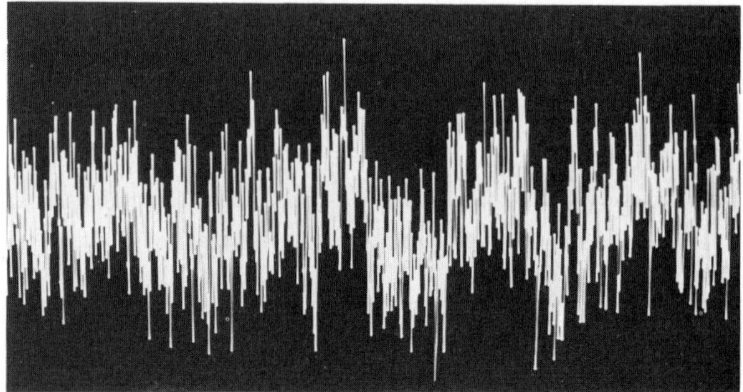

Figure 1.3 Oscilloscope trace for 'white' noise.

The experiment also draws attention to one of the many very remarkable properties of our ear–brain systems. In spite of the differences in the trace, and so, of course, of the different sequences of pressure changes that are being made, our brains recognise the words as being the same! In passing we might note that this is something that digital computers find very difficult. There is no problem in making a computer speak; but to make a computer recognise words spoken by any voice is a more difficult problem.

1.5 PITCH AND FREQUENCY

A sound that resembles that of white noise is applause. And that gives us a clue to the nature of white noise. Applause is obviously a series of clicks, each produced by a single hand clap, following each other in rapid and random succession. What would the sound be like if we made the clicks follow at precisely regular intervals? This of course is the sort of sound we get if we listen to a circular saw, or a piece of card being hit by the spokes of a bicycle wheel. It is a rather raucous sound, but it does seem to have an approximate musical pitch associated with it. The faster the clicks follow each other the higher is the pitch. Careful experiment with an electronic device called a pulse generator attached to a loudspeaker and an oscilloscope will soon show that if the rate at which the pulses follow each other lies between about 30 per second and about 20,000 per second we hear a note. Figure 1.4 shows one such trace. If the rate is less than about 30 per second the sound seems to become more in the nature of a 'feeling' of vibration. If the rate is above about 20,000 per second the mechanism of the ear–brain system is unable to follow and we cease to hear any sound. These lower and higher limits are often called the 'limits of audibility'. They vary a great deal from person to person and, in particular, as we grow older the top limit gradually becomes lower. Many people in their

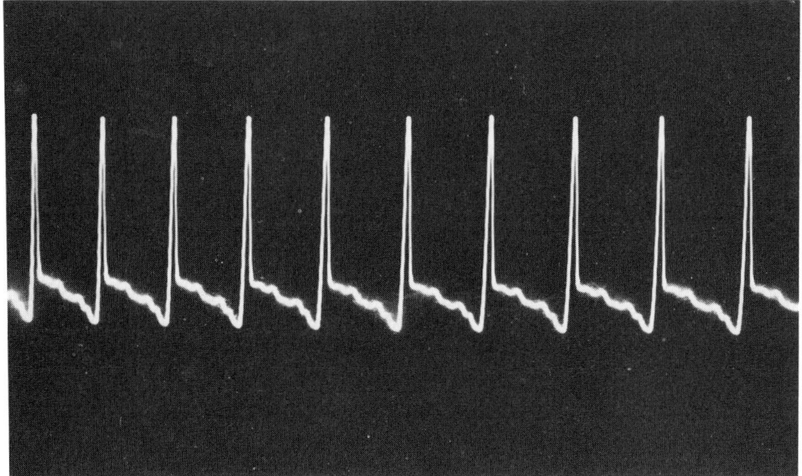

Figure 1.4 Trace for regular pulses.

late sixties or early seventies hear very little above about 10,000 vibrations per second.

One of the many electronic keyboards that are now available (see section 5.8) can be set to produce pulses and it is then easy to show that, each time the musical note goes up one octave, the rate of the pulses is doubled. Equal steps of one octave correspond to successive multiplication by two. But we shall have more to say about that when we talk later about musical scales. The rate per second is usually called the 'frequency' and is measured in Hertz (Hz). 1 Hz means one repetition per second. A frequency meter is useful when demonstrating the relationship between musical pitch and physical frequency. It is important to notice that the relationship between the scientifically measured quantity (frequency) and the musically perceived quantity (pitch) is not always precise (see section 1.7).

There are many systems used by musicians and instrument makers to indicate the musical pitch of a note and this is probably the best point for a brief discussion of them. First of all we need to decide on a standard of pitch so that all instruments can play at the same pitch. This in itself has been the subject of enormous controversy throughout history but the number of different pitches used today is very much less than it was say 100 years ago. The most commonly used international pitch now defines the note A as 440 Hz. But then, of course, there is a whole series of notes an octave apart which are all called A and we need a system to indicate which is intended. Again there are many systems and they are discussed in Appendix B. The one that I shall use throughout this book—largely because it seems to be the most logical and makes the least demands on the memory—is the American Standard system. This starts with the lowest octave on the piano and calls this C_1 to C_2: the A in this octave (labelled

A_1) has a frequency of 55 Hz. The A on which the orchestra tunes (440 Hz) thus becomes A_4.

I should also mention at this point the convention adopted by organ builders for specifying the pitch range of sets of pipes (see also section 4.20). Any set (the organ builder's term is 'rank') of pipes that sounds at the normal pitch, as for example that of a piano, is described as 8 foot pitch, because the lowest open diapason pipe of such a rank would be 8 foot long. A rank of pipes sounding an octave higher would be 4 foot pitch, an octave higher still would be 2 foot pitch. One and two octaves lower would be 16 foot and 32 foot pitch, respectively. Thus, if the key corresponding to the A above middle C, is depressed with an 8 foot pipe in use, the note sounded would be A_4 in American pitch notation and would have a frequency of 440 Hz. But if the pipe were 4 foot, then A_5 would sound, and if a 2 foot pipe were used it would be A_6, and so on.

So-called mutation stops are also in use, to add to the confusion, which sound not octaves above or below the key pressed, but some other interval. Thus a stop of $2\frac{2}{3}$ foot pitch would sound at an interval of a twelfth (an octave plus a fifth) above the note in the normal pitch range.

It is interesting to note that although harpsichords do not use pipes, but strings, nevertheless the pitches of the different stops are still designated by the terms 8 foot, 4 foot, etc.

1.6 DETECTING ULTRASONIC WAVES

Of course we can make microphones that can detect sounds over a very much wider range of frequencies than those that can be detected by the human ear–brain system. The region below the lower limit of human audibility is sometimes called 'subsonic' and the region above the upper limit is usually called 'ultrasonic'. John Tyndall had no microphones at all but he used a special flame that was good at detecting high frequencies. He called it a sensitive flame. It is quite easy to make one.

> A glass tube of about 5 mm in internal bore is heated and drawn out into a tapered jet which is between 0.5 and 1 mm in internal bore. The end must be cut square and then attached to a gas supply. It is important that any turbulence or fluctuation in the gas supply should be eliminated and this is best done by packing the tube loosely with glass wool. The jet is lit and the supply adjusted until a long flame is produced for which a very tiny increase in flow will make the flame turbulent. The adjustment is quite critical, but when properly executed the flame becomes very sensitive to high pitched sounds. Rattling a bunch of keys, speech containing sibilants, and even an ultrasonic dog whistle

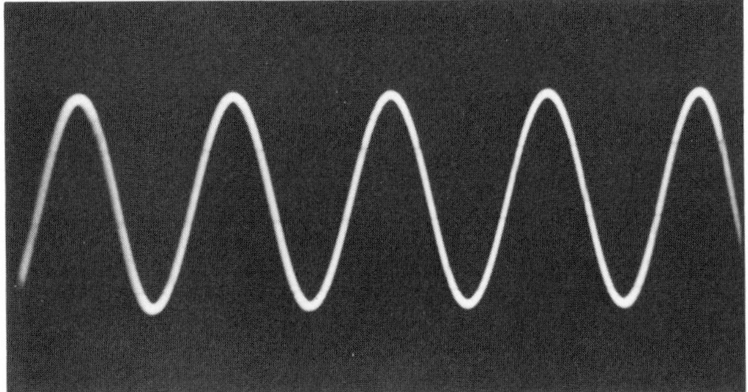

Figure 1.5 Trace for pure tone.

will make the flame duck from the long jet form to the short turbulent form.

Bats use sound at ultrasonic frequencies as a form of radar or sonar for navigation. Each kind of bat uses different frequencies and different techniques so that detecting their navigation pulses is a very good way of identifying bats from a distance. Professor David Pye has developed a series of 'bat detectors' which convert the ultrasonic signals into audible sounds and, using such a detector it is easy to show that the sensitive flame is responding chiefly to the ultrasonic components in the rattle of keys or the sibilants in speech.

1.7 PURE TONES

Though trains of pulses occurring at a frequency that is well within the audible range have an identifiable pitch, they make rather harsh and unpleasant sounds. If we replace the electronically generated succession of pulses by a more mellow sound, for example by the kind of note a musician might call a pure tone (e.g., a flute playing a steady note, quietly, and without vibrato), then the oscillograph trace (figure 1.5) reveals that the pressure is swinging smoothly from high to low and back, rather than changing suddenly as for the pulses. Such a note needs only two numbers to specify it completely. One represents the size of the pressure changes and the other the frequency at which they occur. Mathematicians would describe the wave shape as a sine wave. Again, the musical pitch of the note is roughly related to the frequency; the higher the frequency, the higher the pitch.

From time to time we shall need to talk about sound waves corresponding to pure tones and it might be useful at this point to spend a moment or two defining some of the terms that need to be used. Figure 1.6 shows a drawing of a sine wave; the frequency is the rate at which the pressure goes

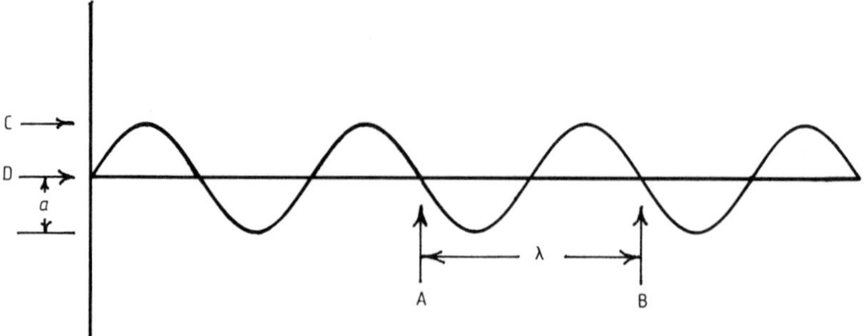

Figure 1.6 Sine wave with parameters.

through its cycle at any one point. (The term 'period' is sometimes used for the time taken to complete one cycle; thus a tone with a frequency of 1000 Hz could be said to have a period of one thousandth of a second, or one millisecond.) For example at point A, as the wave moves along in the direction from left to right the pressure difference from the general level in the room will rise to a maximum value, then fall to zero, go on falling below the room level to its maximum negative value and then rise back to room level again. This represents one cycle and the number of such cycles completed in each second is the frequency (f). The wavelength (λ) is the distance that the wave travels in one cycle, for example from A to B in figure 1.6. The amplitude of the wave is the difference between the pressure at a maximum and that at room level, that is C–D in figure 1.6.

It is very important to be clear that the frequency (the number of complete waves every second) is what you measure scientifically and is completely objective. The musical pitch is what you hear and is quite subjective. In order to stress just how subjective it is, a good demonstration is to feed a sine wave at about 440 Hz via an amplifier to a loudspeaker with the gain set quite low so that you can only just hear it. Then turn the gain up suddenly. The measured quantity (frequency) has not changed, but many people will sense that the pitch changes as well as the loudness.

If this experiment is tried with a large audience it is usually found that about a third of those present hear the note go down in pitch, a third hear it go up, about a quarter don't hear any pitch change, and a few can't make up their minds. The effect is a dynamic one; in other words it only seems to occur when the loudness actually changes. Fortunately too it only seems to occur with pure tones. I say fortunately because obviously, if every time an orchestra played a sudden loud chord, some members of the audience heard it as flat and others as sharp then there would be some confusion. If a much lower frequency than 440 Hz had been chosen then most people would have heard the note go down in pitch (i.e., go flat) and, if the note had been two or three octaves higher, most people would have heard it go up in pitch (i.e., go sharp).

We will return later to some of the fascinating questions that arise when we study the mechanism of hearing in more detail (see sections 1.10 and 1.17). But for the moment we have already reached the point at which we can begin to make a few simple musical instruments.

1.8 MUSIC FROM WOODEN BLOCKS

In the late 19th century Herman von Helmholtz carried out many interesting experiments on musical acoustics and most of his apparatus was made by Rodolf König. König duplicated much of Helmholtz's apparatus for sale and among other things he made sets of wooden blocks that could be used to illustrate some basic principles of musical sound production. The set I use consists of eight blocks of pitch pine. Each is 200 mm long and 30 mm wide and all are of a different thickness. The thinnest is 7 mm and the thickest is 17 mm. When they are dropped on to a hard floor, beginning with the thinnest, a rising diatonic scale is heard (C,D,E,F,G,A,B,C′). What happens is that they usually strike one end first and so bend. The 'springiness' of the wood then makes them straighten out, overshoot, and so on, continuing to vibrate for a few tenths of a second; long enough for the ear to identify the note. Any reasonably hard wood will do, teak is particularly good, but the tuning has to be done by trial and error. I have larger sets consisting of the first ten or twenty notes of a tune. The blocks can be set on the edge of a table and then the tune can be played merely by dropping each block in turn and the player has only to think about the timing. Many other objects will emit notes when dropped; but, of course, the frequency may well be outside the normal hearing range. For example the bat detector that we mentioned in section 1.6 can be used to reveal that even an ordinary pin dropped on the floor is emitting ultrasonic waves and a handful of hardened sewing needles sound like a peal of bells when the frequency is brought into the audible range by the detector.

1.9 THE FIRST FAMILY OF MUSICAL INSTRUMENTS

Obviously the blocks form a very impractical instrument because to repeat the tune the blocks have to be picked up and reset! But, nevertheless, it does illustrate one of the simplest ways of playing tunes on a musical instrument. You simply use a separate vibrating device for each note; I tend to call instruments of this group the first family. The obvious first member is the xylophone which, in its simplest version is just our collection of wooden blocks laid out in order on pads of some kind and struck with a light hammer. In more sophisticated versions there are also resonators...but we

Figure 1.7 A Nigerian 'Sansa', or thumb piano.

shall discuss that modification in the next chapter. There are many other members of this first family, such as, for example, the harp, the celeste, the organ and, of course, the piano. Many non-European countries have instruments of this family. For example, in various parts of Africa 'thumb-pianos' of many different designs are made. Each note is provided by a separate springy-steel strip, often made from a large sacking needle, and the tuning is done by altering the vibrating length. The one illustrated in figure 1.7 is called a 'Sansa' and comes from Nigeria. Although capable of playing a melody, they are usually used to provide a complex rhythm.

It is important to recognise that a piano is really eighty- eight separate instruments, one for each note. The importance of this from the scientific point of view is that when the instrument is being built, the instrument maker can set up each combination of strings, hammer, dampers, levers, keys, etc, to be exactly right for each note. No compromises are needed as is the case in many other instruments. In strings and woodwind, for example, each note is produced by modifying in some way the same vibrating system; it cannot therefore be exactly right for each note. This means that there must be compromises, but it also means that the tone quality from note to note is more likely to differ than with instruments in the first family. We shall discuss instruments of families other than the first in Chapter 2.

1.10 THE EAR AND HEARING

In section 1.8 we discussed the sounds produced by dropping wooden blocks on the floor and we mentioned that we only hear a musical note if the frequency of the vibrations is within the range of the human ear. It is also important that the material is not so soft that the vibrations die away almost instantly so that the ear–brain system has no time to identify a

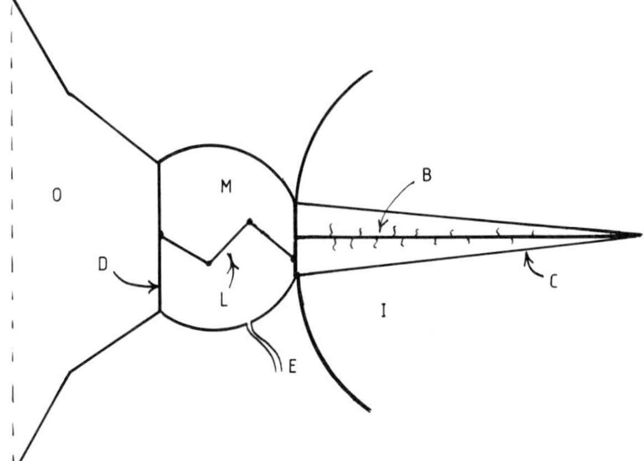

Figure 1.8 Diagrammatic representation of a human ear.

note. This might therefore be a good time to discuss the mechanism of hearing and in particular how our sensitivity varies with frequency.

Often, especially in children's books, the ear is represented as a telephone receiver which passes signals on to a telephone exchange in the brain. I suppose this is valid in the sense that the ear has a diaphragm and converts incoming sounds into a signal which is sent along the auditory nerve to the brain. But there are many very significant differences, perhaps the most important of which is that a great deal of processing of the signal occurs between the diaphragm and the brain itself. It is the enormously complex interactions between the diaphragm and the brain that provide the miracle of hearing.

We shall first consider the ear itself as a piece of physical apparatus, rather then venturing into anatomical or physiological discussions. In order to underline this point we have deliberately omitted the well known diagram of the anatomy of the ear which appears with monotonous regularity in textbooks on sound. Figure 1.8 is our substitute. There are three obvious divisions of the ear in most mammals. First the outer ear (O), then the middle ear (M) and finally the inner ear (I). The outer ear, at first glance, appears to be merely a funnel to collect the disturbances in the air. It amplifies the sound simply by channelling the energy dispersed over a relatively large area down a tube and concentrating it onto the relatively small diaphragm or drum stretched across the end of the tube; the old fashioned ear trumpet is an extension of the same principle to increase the amplification still further.

But there are other important functions related to directional sensitivity. Even in humans, turning the head can help us to locate a sound. But watch the very much larger and very mobile ears of a horse or of a German shepherd dog and their rotation to lock on to the direction of a sound is very obvious. In humans the shape of the outer ear, or pinna, plays a further part in enhancing our stereo sense. A sound wave impinging on the head say from the right is modified by diffraction and the wave that reaches the pinna of the left ear is quite different from that reaching the right. In particular the frequency distribution is different. But the whole ear–brain system is designed to interpret these differences in order to locate the source of the sound. The most striking demonstration of this is the use of so-called 'dummy-head' recording. A model of a human head of the kind used on shop-window dummies has a microphone inserted at the bottom of a hole on each side of the head, drilled to represent the ear canals. If this is used as a stereo pair of microphones the fidelity of the resultant recording is startling. The resulting recording needs to be heard through head phones, but the realism of the directional information far outweighs anything that can be done with an ordinary stereo-pair of microphones.

The ear drum (D), or tympanum, is under involuntary control by muscles that can change the tension. This seems to be a protective device to minimise damage under conditions of extreme loudness or external pressure change rather than an essential part of the hearing mechanism. The pressure changes in the incoming sound wave are followed by the drum and are transmitted through the middle ear via a set of three linked bones (L) (the malleus, incus and stapes, *or* hammer, anvil and stirrup) which act as a lever system to change the relatively large displacements of the drum into much smaller, but more powerful movements of the oval window which transmits the vibrations to the inner ear. The bone linkage acts in much the same way as a matching transformer in the output of an audio amplifier.

The middle ear is a closed chamber and hence variations in the total atmospheric pressure would cause distortion of the drum and could result in damage. But a leakage path is provided, called the Eustachian tube (labelled E in figure 1.8), which leads to the back of the nose and allows the pressure in the chamber to be equalised with that outside the drum. Blockage of this tube, for example by infection, can cause the pressure equalisation to take a longer time, or to cease, and the result is partial deafness and discomfort. Very sudden changes in the external pressure such as can be caused by a train suddenly entering a tunnel, or sudden changes in height in an aircraft can cause temporary effects of the same kind.

The third division, the inner ear, is where the vibrations of the oval window are fed into the liquid contained in the cochlea (C) which then affects the multitude of fine hairs—some 25,000 in humans—the movement of which in the fluid produces the signals that eventually reach the brain through the auditory nerve. These hairs are mounted on the basilar membrane (B) which divides the cochlea down the middle. Though the cochlea is shown straight in the diagram, it is, in reality, coiled into a spiral.

Early theories suggested that each hair responded by simple resonance to different frequency components in the incoming sound. This is clearly inadequate to explain many of the miraculous properties of our hearing system and there is still controversy about the precise mechanism by which the waves in the cochlean liquid become signals to the brain. One fact that has become clear through animal experiments is that the signals passing along the auditory nerve still retain some of the characteristics of the original sound waves and are not merely coded information about frequency components. Another fact is that there seems to be a kind of 'pre-sorting' carried out by the cochlea; incoming signals are in effect sorted into frequency ranges called 'critical bands'. One of the ways of demonstrating this is to study the interference, or masking, that occurs between sounds of different frequencies. Sounds within a given critical band will interfere, but sounds in separate bands can be heard together. However, fascinating though it may be, we shall not describe the mechanism in detail; what is more important to us at this stage is the variation in sensitivity of the complete hearing system.

1.11 MEASUREMENTS ON HEARING

In terms of amplitude, the so-called threshold of hearing, the tiniest sound that we can just hear in conditions of absolute silence, represents a change in pressure of something like 2 parts in 10,000,000,000 of the atmospheric pressure. At the other end of the range, a sound that just ceases to be heard normally and becomes a sensation of pain corresponds to a change a million times larger. In absolute terms the threshold of hearing is round about 20 millionths of a Newton per square metre and the threshold of pain is 20 Newtons per square metre. (Atmospheric pressure is about 100,000 Newtons per square metre.)

Of course these figures can only be taken as rough guides. The sensitivity changes from person to person, it varies with age, it varies with the frequency and with the type of sound and also with the surrounding conditions under which the test is made.

A great deal of information is now available on the kinds and levels of sound that can cause discomfort or even permanent damage to the hearing system and we shall discuss some of the practical implications in section 6.13.

The measurement of sound levels is a difficult problem primarily because we are trying to find a purely physical way of measuring a response that is psycho–physical. Since this is a manifestation of some of the general principles involved in relating subjective sensations to objective measurements we will spend a little time discussing the problems.

The earliest work seems to have been done by E H Weber early in the nineteenth century. He worked with weights and the sensation of touch. An observer was blindfolded and various weights were placed on the outstretched palm of one hand. It was found that when small weights only were in place a very small increase in weight could be detected. However, if a large weight was already in place, a very much greater increase was needed before the change was detected by the observer.

In crude terms one can convey the sense of his discovery by saying that if you stick a pin into someone, the first thousandth of an inch is quite painful; but if the pin is already in half an inch, another eighth of an inch makes little difference!

In the realms of vision and of hearing Weber's observations can be made much more precise and ultimately a law (the Weber–Fechner Law) emerged which can be stated as:-

> The increase in stimulus (the physical quantity) needed to produce a given increase in sensation (the perceived quantity) is proportional to the pre-existing stimulus.

This means that, for example, if we listen to a sound which is fairly quiet and we then double the intensity of the sound (the physical quantity) we can detect a change. But if we then listen to a sound whose intensity is twenty times that of the first, we shall need to add twenty times the intensity we added to the first sound if we are to experience the same apparent perceived increase.

The law is not precisely obeyed by any of our sensations, but there are many examples of everyday experience that confirm the general idea. For example at night when the general light level is very low, motor car headlights can be absolutely dazzling; but in day time when the ambient light level is high, one can stare right into the headlights with no discomfort at all. And the well known teacher's ploy of persuading a class to be quiet by trying to hear a pin drop obviously has a scientific basis.

In mathematical terms this kind of law leads to the use of logarithmic relationships. We have already met this in section 1.5 where we found that each time we add the interval of one octave we have to multiply the frequency by two. So important is this idea that, in spite of my aim to keep out mathematics as much as possible, a little has to come in here. Everyone nowadays has heard of 'decibels' in connection with the measurement of noise levels but many people do not really understand the term.

The first point to get clear is that the decibel itself is not a measure of loudness, indeed it is not a unit of anything; it is concerned with the ratio of two quantities. When one reads that some source of noise such as a pneumatic drill has a noise level of 90 decibels, what is implicit is that its level is 90 decibels above the threshold of hearing. The omission of the last five words is what often leads to confusion.

But let us start a little further back. Figure 1.9 shows on the left a

Threshold of pain	120 db	— 1 watt/metre^2
Riveting steel plates about 6 feet away	110 db	— 100 milliwatts/m^2
Large symphony orchestra playing fff in a concert hall	100 db	— 10 milliwatts/m^2
A pneumatic drill at about 10 feet away	90 db	— 1 milliwatt/m^2
	80 db	— 100 microwatts/m^2
Traffic in a busy city street	70 db	— 10 microwatts/in^2
A crowded restaurant	60 db	— 1 microwatt/m^2
	50 db	— 0.1 microwatt/m^2
Background noise in a town at night	40 db	— 0.01 microwatt/m^2
	30 db	— 0.001 microwatt/m^2
	20 db	— 0.0001 microwatt/m^2
A mosquito passing one's ear	10 db	— 0.00001 microwatt/m^2
Threshold of hearing	00 db	— 0.000001 microwatt/m^2

$$= \frac{1}{1{,}000{,}000{,}000{,}000} \text{ watt/m}^2$$

Figure 1.9 List of common sounds arranged in approximately equal steps of increasing loudness with the corresponding level and sound intensity.

series of common sounds that has been arranged vertically in what seem to the average ear–brain system to be approximately equal steps of increasing loudness. In the centre column are the increases in level expressed in decibels (dB). But on the right are given the corresponding sound intensities measured in watts per square metre, that is the rate at which sound energy is falling on the ear or the measuring system. The important point to notice is that each increment of 10 dB in the middle scale corresponds to multiplying the intensity by 10. (The mathematical relationship is that the number of decibels is ten times the logarithm of the ratio of the intensities. So, for example, doubling the intensity results in an addition of $10 \times \log 2$ decibels which is 10×0.3010 which is 3 dB.)

Figure 1.10 shows how the amplitude of the pressure change in a given

Amplitude of pressure change in the wave		*Sound pressure level*	*Intensity (energy flow in unit time through unit area)*
Newtons per square metre	*Millionths of an Atmosphere*	*Decibels* (dB)	*Millionths of a watt per square metre*
20	200	120	1,000,000
2	20	100	10,000
0.2	2	80	100
0.02	0.2	60	1
0.002	0.02	40	0.01
0.0002	0.002	20	0.0001
0.00002	0.0002	0	0.000001

Figure 1.10 Further comparisons of sound levels and intensities.

wave can be expressed in Newtons per square metre or in millionths of an atmosphere and compares this with decibel ratios and with intensity measured in millionths of a watt per square metre.

This may sound neat and convincing but, unfortunately, it is a considerable over-simplification. We have not the space, nor would it be appropriate to delve very deeply into the real complications, but we do need to proceed with the arguments a little further. It is important to realise that our hearing sensitivity varies enormously with frequency, and with the kind of noise being measured. One of the most serious problems is the one that arises if, when measuring sound by comparing with a standard sound, the two sounds have very different qualities. For example, a comparison between a pure tone and a regular succession of pulses, or between a note on a violin and that of a trumpet could present considerable difficulty.

To take care of this last problem, or at least to ensure that everyone knows what kind of measurement has been used we need to make some definitions.

If we listen alternately to the sound under measurement and to a standard sound of exactly the same quality then when the two match in subjective loudness the level of the standard sound in dB above the threshold for the same quality of sound is described as the 'sensation level'.

If, on the other hand we compare the sound being measured with a pure tone of frequency 1000 Hz, then when they appear to the ear–brain system to be equally loud, the ratio expressed in dB of the 1000 Hz tone to the threshold for 1000 Hz, is called the equivalent loudness of the non-standard tone and is measured in 'phons'.

There are immense problems in making measurements on the hearing system, not least because of the way in which our brains operate. It turns

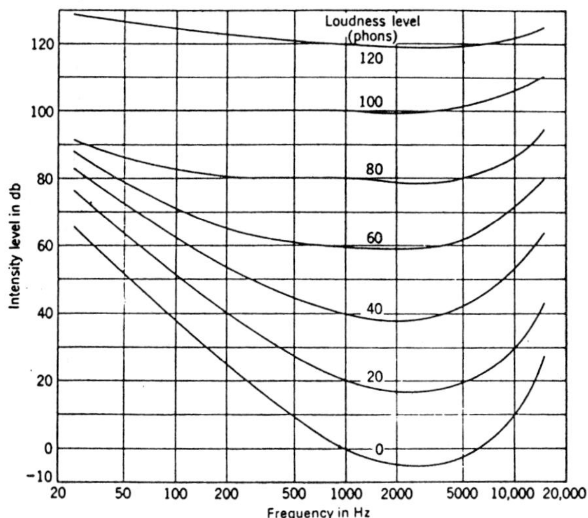

Figure 1.11 Fletcher–Munson lines of equal subjective loudness.

out that when you attempt to do an experiment involving someone's ear–brain system the very act of making the experiment changes the way the subject responds to later experiments. This problem is discussed in section 1.17. Another problem is in comparing tones of different frequencies. A large number of observers is asked to listen to a whole series of pairs of tones in which the two notes have different frequencies. They hear each note of a pair alternately and are asked to adjust the tones to be equally loud. This experiment leads to a well-known set of curves usually known as the Fletcher–Munson equal loudness curves. Figure 1.11 shows a typical set. You will see that along the vertical line corresponding to a frequency of 1000 Hz the curves are labelled 0, 20, 40, 60 dB. These are the decibel equivalents of the ratios of the intensity levels of a series of 1000 Hz tones of increasing loudness with respect to the threshold of audibility at 1000 Hz. We can therefore call these loudness levels in phons. But if we tried to compare a tone of say 200 Hz whose actual intensity level is 40 dB, with the 1000 Hz tone we should find that they would only sound to be equal in loudness if the 1000 Hz tone were at 20 dB. The curves join up the points at which the subjective loudness is the same and so the values on these curves can be taken to be equivalent loudness levels in phons.

One interesting point that emerges, which is of particular significance to Hi-Fi enthusiasts, is that for very loud sounds the variation between the physically measured intensity level and the subjectively judged equivalent loudness does not vary greatly with frequency, i.e., the curves are almost horizontal straight lines. But at low levels, such as 20 or 40 dB, there is an enormous variation. This explains why many Hi-Fi amplifiers have a 'loudness' control. This increases the amplification both at low and high frequencies relative to the middle so that the resultant sound has approximately equal equivalent loudness over the range. A uniform amplification would result in loss of bass and treble whenever the overall level is low.

Even though these curves are related to our subjective sensations, they are, nevertheless, based on physical measurements of intensity levels. Thus the loudness level all along the curve labelled 60 dB is 60 phons, and that of the next curve up is 80 phons. The ratio of their intensity levels at 1000 Hz is thus 20 dB, or 100:1 and we can measure this on a meter. But if we listen to the two tones at 60 and 80 dB does one sound 100 times louder than the other? The answer is certainly 'No'. Many attempts have been made to try to establish a means of assessing this subjective loudness. For example an observer might be asked to compare a given tone heard by one ear with that of a quieter tone heard by both ears and it might be reasonable to say that if the result is the same loudness for both, then the tone heard by one ear must be twice as loud as that heard by both. Another method has been to use two tones of different frequency set to equal loudness and then to use the two together as an approximation to a tone that is twice as loud as either. My intention is not to get deeply involved with the details of the various methods, but rather to show, on the one hand, how difficult this kind of measurement is, and, on the other hand that there are methods of attacking these difficult problems and that, in the end, it is surprising how consistent a large series of measurements can be.

There is one final unit that should be mentioned. It is called a 'sone'. As a result of many thousands of measurements with many different observers, a composite curve (figure 1.12) can be produced which shows the relationship between the loudness level as measured in phons and the subjectively judged loudness in sone. Quite arbitrarily a tone of equivalent loudness level of 40 phons is defined as 1 sone and you will notice that the curve passes through the point corresponding to 40 phons and 1 sone. A sound of subjective loudness 2 sones would then seem to an observer to be twice as loud, but from the curve we can see that its equivalent loudness as measured would be 49 phons. For very loud sounds the subjective loudness in sones seems to double for every 9 phons increase in measured equivalent loudness. For quiet sounds the subjective loudness increases much more steeply.

1.12 MORE QUESTIONS ABOUT THE NATURE OF MUSIC

So far we have discussed the nature of single musical notes and how they differ from noise but we are obviously still a long way from being able to answer the main question of this chapter. What is music?

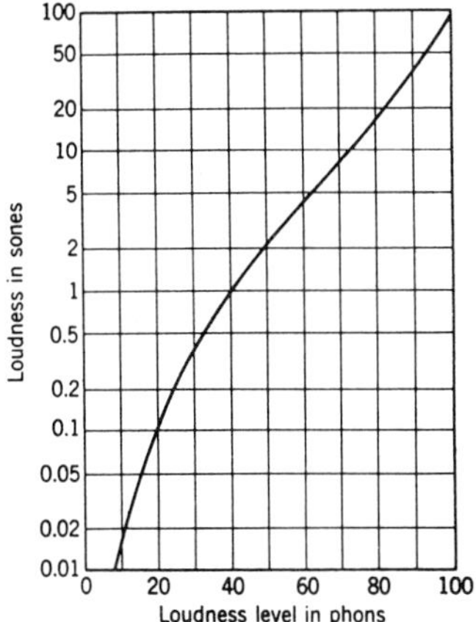

Figure 1.12 Relationships between loudness level in phons and subjective loudness level in sones.

Look at the traces of figure 1.13(*a*), (*b*), and (*c*). One is the sound of an audience entering a lecture theatre to hear a lecture. One is the sound of a symphony orchestra at the end of the first movement of Mendelssohn's violin concerto and the third is the sound of a symphony orchestra tuning up before a concert. They are arranged in random order. Can you tell which is which? They all look much the same, and, indeed they all superficially resemble the trace of figure 1.3, white noise. But obviously they are very different in terms of the information that they are conveying. It is interesting just to reflect for a moment on the fact that when we listen to any one of these sounds we are receiving information that is extremely precise. In a second or two we have recognised the kind of sound; if it is music we have recognised the kind of music and possibly even recognised the composer and the piece. When the same information is presented to an audience on the cathode ray tube without the sound, the audience is receiving precisely the same information but it is difficult, if not impossible, even to identify the kind of sound.

The answer to the identities of figure 1.13(*a*), (*b*), and (*c*) is given at the end of the chapter. I have to admit that I find this kind of test very difficult and, even though I have been looking at some of these traces for many years, I still cannot identify them with any certainty.

One of the traces is for a symphony orchestra tuning up and this, at once, highlights a major problem. It is music in the sense that it is sound made by musical instruments; but it is noise in the sense that no one would wish to listen to it for more than a few minutes at the beginning of a concert to whet the appetite.

Before we leave figure 1.13 there is another important point to be made. You will notice that each trace consists of a single line. This is to be

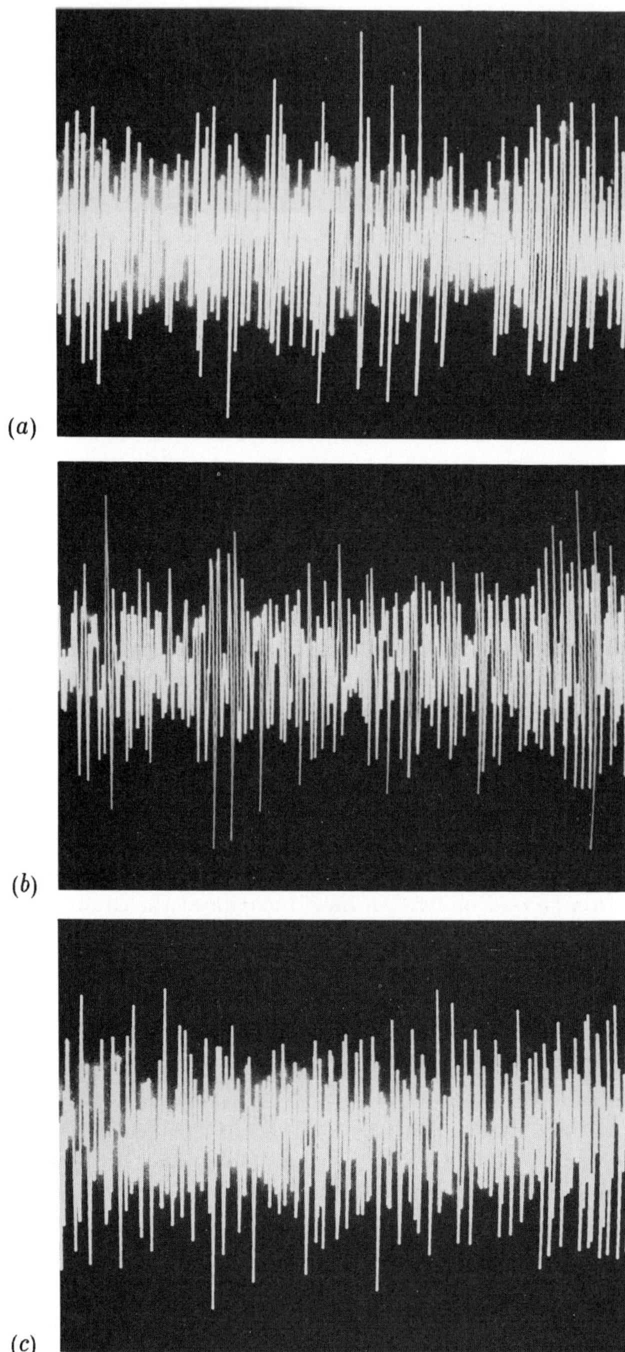

Figure 1.13 Traces of three different sounds. Can you identify them? Answers at the end of Chapter 1.

expected because it represents the changing pressure just in front of the microphone that recorded the sound. Pressure can only have one value at one point at a given moment in time, though it may change very rapidly with time. Yet when we listen to the sounds corresponding to these traces it is immediately obvious that there is more than one source of sound. In the recording of the audience entering the theatre you can hear all sorts of separate coughs, giggles, rattles of toffee papers, snatches of conversations, etc. And in the orchestral examples you can pick out what each instrument is doing; with a little practice you can listen to the violins, or the clarinets, the trumpets, etc. One of the problems that we shall come back to later is just how the human brain can do this. A man-made computer would find it very difficult to analyse one of the traces of figure 1.13 unless it was told in advance what instruments were present. But the marvellous human brain not only performs the feat but does it incredibly rapidly.

It is quite clear that, although our ear–brain system can distinguish immediately between music that we like and noises that we do not like, the oscillograph trace really does not help very much. One of the problems is that the traces of figure 1.13 correspond to complex mixtures of sounds. Usually, at the beginning of a piece of research a scientist would start with very simple systems in which the number of variable quantities has been reduced as much as possible. Then as the research progresses it is important to reintroduce the variables gradually until the whole complex reality of the system is understood.

It is an extraordinary fact that, in the field of musical physics, this point has often been missed. Scientists studied one note of an instrument being played artificially in the lab and extrapolated the results to suggest that they understood what was happening in a real instrument played by a real musician in concert hall conditions. Fortunately over the last twenty or thirty years the importance of the real conditions has been increasingly realised.

1.13 MUSIC AND INFORMATION

Nevertheless let us go back to some simple sounds and look for more clues about the nature of music. In section 1.4 I mentioned the experiment of varying the loudness of white noise and adding a little seagull noise. Immediately the feeling of being at the seaside was invoked. In other words the sound was conveying information. A steady pure tone is very dull, but by varying it, e.g., the wolf whistle, some kind of information is imparted. In fact it is quite surprising how much information can be transmitted with even one note, if it is interrupted. That in fact is exactly what Morse code is. A single note that stops and starts to give 'dots' and 'dashes' can be made to convey whole messages. The talking drums or trumpets of Africa can transmit information on two notes.

The system is very ingenious. Words in any language (including English) can be imitated by sequences of just two notes (say, for example, C and E). For example 'father' and 'mother' could both be represented as E C. But grandfather would be E C C. How does one eliminate the large number of words that would be represented by the same note sequence? It is done by using standard phrases. For example 'mother' might be replaced by the phrase 'mother, who bears the children' and father by 'father who hunts for food'. Very quickly then the sequences of notes take on a unique meaning.

The possibility of conveying vast amounts of information by a simple sequence of sounds is demonstrated very vividly by the *Directory of Tunes* published some years ago by Denys Parsons. I confess that when I was first told about it I did not believe it. But I am now an enthusiastic convert and would not be without the book. How many times have you found yourself listening to, or perhaps even humming, a tune that you know very well but cannot remember either its title or who composed it. In the directory all you need to know is the sequence of the first fifteen or so notes, not in musical notation, nor even in any particular key. All you need to know is whether each successive note goes up, goes down, or is repeated. For example the tune of God save the Queen would be written out as follows

*RUDUU URUDD DUDDU

The first asterisk represents the first note, U represents Up, D, Down and R, Repeat.

Incredibly this is enough to identify some 10,000 classical themes and 6000 popular ones. The division into groups of five symbols simply makes it easier to find the item in the Directory in which the entries are arranged alphabetically.

Figure 1.14 shows two typical extracts which should illustrate the way in which the system works. But apart from extolling the virtues of the book my main purpose was to underline the whole idea that music is concerned with information.

1.14 WHAT CAN BE VARIED?

What are the factors that can be varied in order to convey information through musical sounds? The basic two elements of a simple tune are rhythm and melody. In other words time intervals and pitch intervals. We have seen in Denys Parson's book that even quite rough ideas of the sequence of pitches can be enough to identify a tune. It is interesting to speculate on whether rhythm alone would be enough. I suspect that it would be much more difficult—though tapping out a rhythm was once used to test a panel on a radio 'quiz' show about music. In the printed account

*DUDUU	DDUUD	DUDUU	**Stravinsky** Petrushka: Dance of the gypsies
*DUDUU	DDUUD	DUUDU	**Stravinsky** symphony in 3 movements 2m 3t
*DUDUU	DDUUD	UDUDU	**Handel** harpsichord suite/5 in E 3m
*DUDUU	DDUUD	UDUUD	**D'Indy** Symphonie sur un chant montagnard 3m 1t
*DUDUU	DDUUD	UUUDU	**Schubert** symphony/8 Bmi 'unfinished' 1m 2t D759
*DUDUU	DDUUR	DRDRD	**Schubert** piano sonata in A 2m 1t D959
*DUDUU	DRDUD	UUDUD	**Mahler** Lieder & Gesänge aus der Jugendzeit/2 Erin- [nerung
*DUDUU	DRDUU	DRUDU	**Mozart** flute concerto/2 D (oboe C) K314 2m
*DUDUU	DRDUU	RDDUD	**Sibelius** Return of Lemminkainen op27/4 3t
*DUDUU	DRRDU	DUUDR	**Beethoven** wind octet in E♭ op103 allegro

*RRRRR	RRRRR	RRURU	**Sullivan** The lost chord (song)
*RRRRR	RRRRR	RRUUU	**Sibelius** Valse triste, orch op44 2t
*RRRRR	RRRRR	RUDDD	**Stravinsky** Petrushka: Tableau 2t
*RRRRR	RRRRR	RUDDD	**Beethoven** symphony/5 in Cmi 3m 2t
*RRRRR	RRRRR	RUDDD	**Hummel** trumpet concerto in E 3m
*RRRRR	RRRRR	RUDRD	**Verdi** Requiem: Requiem aeternam
*RRRRR	RRRRR	RUUDR	**Berlioz** Requiem/8 Hostias
*RRRRR	RRRRR	RUURD	**J Strauss Jr** Tales of the Vienna Woods/5 1t
*RRRRR	RRRRR	RUURR	**Beethoven** symphony/7 in A 2m 1t
*RRRRR	RRRRR	UDDRD	**Mozart** Serenade in E♭ K375 1m 1t

Figure 1.14 Two extracts from Denys Parson's *Directory of Tunes.*

of my 1971 Christmas lectures I included a reference to an attempt by a writer to describe a piece of music in words as follows:-

> Stan Barstow, in his novel *The Watchers on the Shore*, produces a remarkable verbal description of a piece of music. It occurs when Albert has just returned to his flat and joined Conroy and they decide to listen to a record of the Roman Carnival overture by Berlioz.
>
> '...and climbs to a climax that's all snapping, snarling brass...doo-ah rratatah dee doo doh dah, rrumdidumdidum-didumdidumdumdah dooooh daaaAH!'
>
> If you know the overture, this is a brilliant and precise evocation; if you do not, then it could mean all kinds of different things.

But, in addition to rhythm and melody there are other factors such as harmony, orchestration (i.e., the choice of instruments and the distribution of the various components between them), etc. Harmony alone can convey a great deal of atmosphere without actual melody. Consider the opening few bars of the second movement of Beethoven's 7th Symphony. It would be rather meaningless to whistle the theme but figure 1.14(b) shows the appropriate section of the directory of tunes in which the repetition of the first 12 notes is very clear; however, because of the changes in the rich harmony, the effect is strikingly beautiful.

Orchestration can completely change the character of a piece of music as you can soon demonstrate by listening to someone playing an 'Easy arrangement for piano' of a complex orchestral piece. And in one sense a

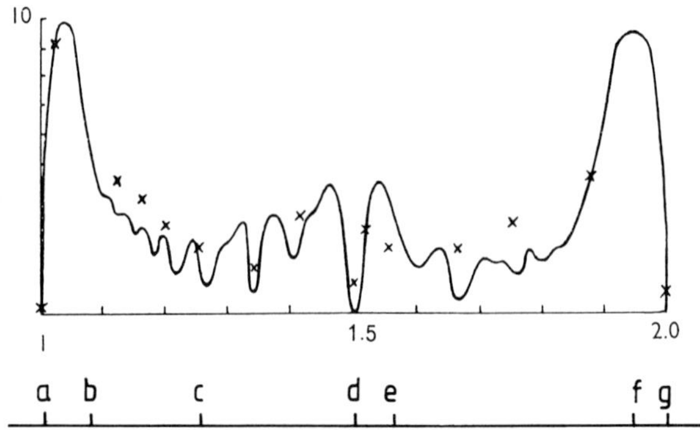

Figure 1.15 Helmholtz's 'degree of roughness curve'. The letters relate to the corresponding traces on figure 1.16.

great deal of this book is concerned with the quality of the sounds produced by various instruments and how they blend together, which, of course, is precisely what orchestration is about.

1.15 HARMONY AND DISCORD

It will be useful at this point if we begin to consider why some pairs of notes sound pleasant together and others unpleasant. Helmholtz made a theoretical study of this using an imaginary experiment with two violinists. One was asked to play a single steady note and the other was asked to start in tune with the first and then slowly glide up through an octave. Helmholtz calculated the 'degree of roughness' to be expected as the experiment proceeded and figure 1.15 shows his result. The quantity represented vertically is the degree of roughness on an arbitrary scale of 0 to 10, 0 being completely smooth. The crosses represent the result of an experiment in which a class of students was asked to estimate the degree of roughness on the same scale. The letters along the baseline are mine and relate to figure 1.16.

Figure 1.16 shows a set of oscillograph traces for a series of pairs of pure tones corresponding to the positions marked by the letters on figure 1.15. Even without listening to the pairs of notes it is immediately obvious that pairs which Helmholtz's curve would indicate to be 'smoother' have far simpler oscillograph traces than those indicated as 'rougher'.

Is it possible therefore that the brain finds it easier to deal with waves that repeat exactly after a very short interval of time and finds difficulties with those that only repeat after quite a long interval? For example a pair of notes that are a perfect musical fifth apart, with a frequency ratio of 2:3, would give wave forms like those of figure 1.16 curve d. But the waveforms

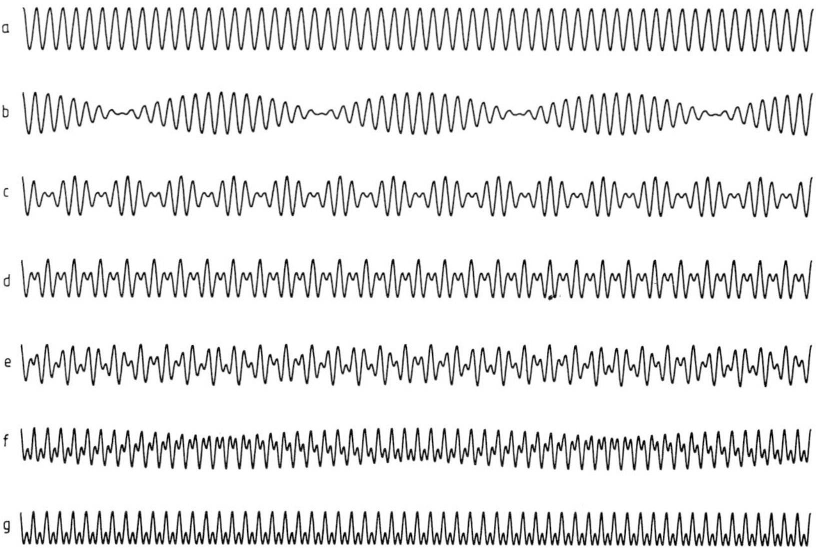

Figure 1.16 Traces of seven pairs of simultaneously sounded pure tones. Their frequency ratios are: (*a*) 1:1, (*b*) 15:16, (*c*) 4:5, (*d*) 2:3, (*e*) 20:31, (*f*) 30:59, (*g*) 1:2.

for a pair that are a musical semitone apart, with frequency ratio 15:16, would give waveforms like those of figure 1.16 curve b. Waveform b takes about seven times longer to repeat than does waveform d.

If two notes are very close together in pitch the loudness 'comes and goes' with the effect called 'beats'. Figure 1.17 shows a model (made by Peter Whye) that can be used for adding two wave forms. In figure 1.17(*a*) the two parts are shown separately and you can see that the shape of the template is the same as the outline formed by the tops of the rods. In figure 1.17(*b*) the template has been slid in and is in such a position that the two waves cancel out. In figure 1.17(*c*) the template has been moved along by half a wavelength and the waves add up to give a wave of twice the amplitude. The effect is called interference—though strictly it should be called non-interference because there is no permanent effect on the two waves. I have sometimes compared the effect to that of someone walking up an escalator that is going down. If the speeds are exactly equal then someone else viewing the person from the side so that only the top part of the body is seen would find it difficult to distinguish the person walking on the moving escalator from someone standing still on a stationary escalator.

Now look at figure 1.18. The model is the same but now the template has a wavelength that is a little shorter than that represented by the rods. Now when they are added the amplitude rises and falls. In figure 1.18(*b*) the waves add in step on the left, but are becoming further out of step as we

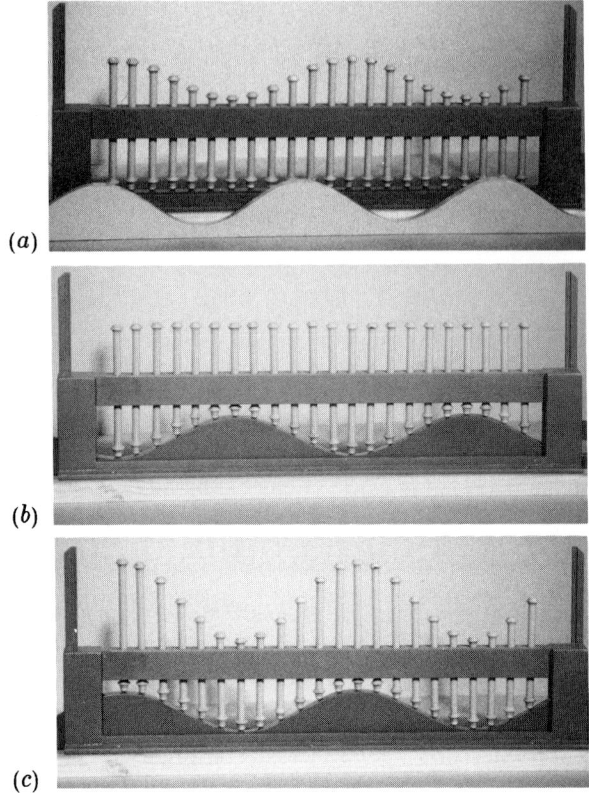

Figure 1.17 (*a*) Components of a wave model to demonstrate interference by addition of waves of identical wavelength. (*b*) Components added out of step. (*c*) Components added in step.

move to the right. In figure 1.18(*c*) the 'in-step' section has moved to the right. This is the beat effect. The easiest way to demonstrate it is to use two identical recorders or tin whistles. Place them both in the mouth at once and blow the same note on each simultaneously. Then place a finger partly over the next lower hole on one. With practice the note of one can be lowered only a tiny fraction and you will then hear the beats quite convincingly. The beats will occur at a frequency that is the difference of the frequencies of the separate notes. Thus if one has a frequency of 880 Hz and the other 873, you will hear 7 beats per second. This effect is sometimes used with organ pipes to give a vibrato effect. A pipe that is fractionally out of tune is called a 'Voix Celeste' and when played together with a similar pipe that is in tune, the vibrato effect is produced (see also section 4.20).

Odd things can happen when you mix sounds. If you look carefully at

(a) 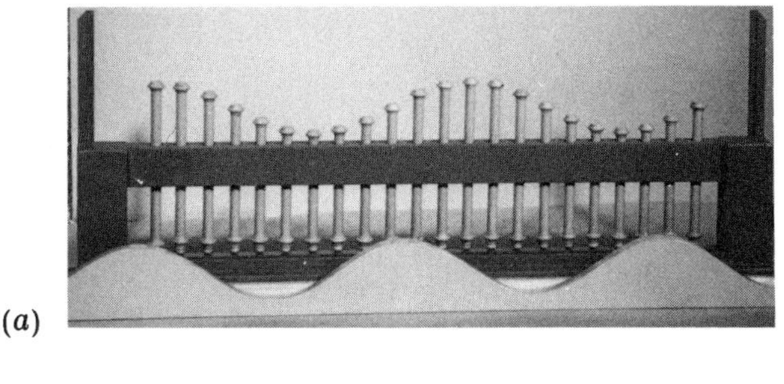

(b) 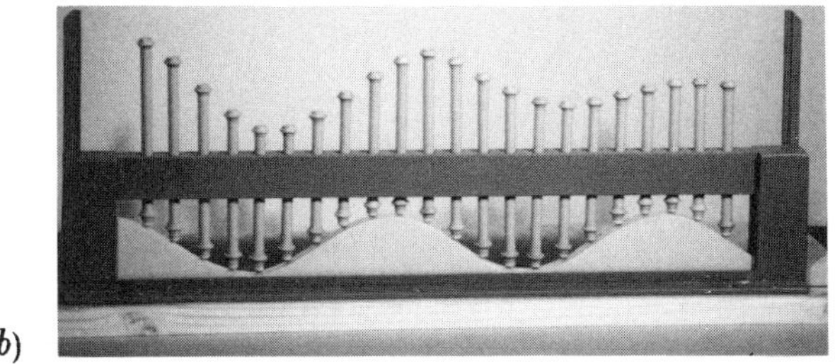

(c) 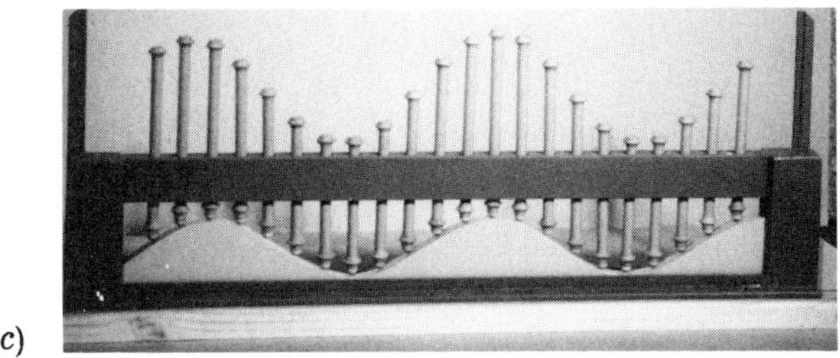

Figure 1.18 (*a*) Components of wave model to demonstrate addition of waves of slightly different wavelength. (*b*) Waves in step at the left. (*c*) Waves in step near the middle.

the kind of whistle that is provided with a life jacket or used to give an alarm you will see that it is really two whistles placed back to back. If you cover the opening of one of the pair while you blow it and then the other you can hear the two separate notes. They are slightly different. But when you blow the whistle normally you do not only hear the two separate notes, but you also hear a much lower pitched note. The lower pitched note can be heard over a greater distance, but to make a single whistle that would give the low note directly would lead to a very long and clumsy whistle. The double whistle is much smaller and lighter but gives the same result.

The low note is sometimes called a 'difference tone' because, for example, if one note had a frequency of 880 Hz and the other 1000 Hz, the low note you hear would have a frequency of 120 Hz, that is 1000 minus 880. The same principle is sometimes used in organs. To produce some of the very low notes, pipes that are 20 to 30 feet long would be needed and it may be difficult to fit these in to the building. Two much shorter pipes placed back to back like the whistles can give the effect of a very much longer pipe; such systems are sometimes called 'resultant bass' (see also section 4.19).

Is the difference tone related to beats? The answer is yes and no! At one time they were thought to be different effects; later they were seen as aspects of the same effect, and nowadays there is a complex theory surrounding the whole phenomenon. We shall give here only a rather superficial view.

1.16 BEATS AND DIFFERENCE TONES

Let us imagine two notes, one of frequency 440 Hz and the other of frequency 445 Hz being played together. As we saw in the last section we should hear 5 beats every second. If we now gradually increase the frequency of the upper note the frequency of the beats will increase. Eventually, however, there will come a point at which the beats are so fast that we no longer notice them. The effect is parallel to that of the flicker of a light source. Fluorescent lights flicker at the rate of 100 Hz when powered by AC mains with a frequency of 50 Hz, but we do not normally notice the flicker. (The flicker is at 100 Hz because the light is bright at the peak of both positive and negative swings of the voltage.) The question of why we do not notice the flicker is an interesting one. Experiments using an electroencephalograph (an instrument that picks up the electrical activity of the brain by means of electrodes attached at various places on the head) show that some evidence of the flicker can be detected in the brain waves themselves even when the frequency of the flicker is well above the point at which the eye–brain system no longer notices it. This suggests that the eye–brain system is quite capable of detecting the flicker, but the mechanism of perception in the brain decides to ignore it as not being of any use.

Something very similar seems to occur with the ear–brain system and, as the difference in frequencies increases there comes a point at which we cease to notice the beats. As the difference in the frequencies becomes higher still (to be precise when the difference in the frequencies comes into the audio range) an apparently different phenomenon begins to occur—the difference tone already mentioned. The early theory suggested that it was due to the non-linearity of parts of the hearing system. A non-linear system is one in which a force applied in one direction will produce a different magnitude of displacement from that produced by a force applied in the opposite direction. This is the same principle that was applied in the early days of radio; the radio frequency waves were amplitude modulated with the signals conveying information but, because the modulation was symmetrical in the positive and negative directions, it balanced out and nothing was heard. However, by introducing a non-linear device, at first a 'crystal' and later a thermionic valve, the modulation could be detected.

In the hearing process the non-linear behaviour of parts of the system enables us to detect the modulation, which in the case of two tones of different frequency is, in fact, the difference tone. The non-linearity is greater when the amplitude is larger and so the effects to be described would appear to occur only when the sounds used are loud. Thus if we listened to two tones, say 400 and 600 Hz, the beats would be at 200 Hz, that is too rapid to be heard as beats, but we should hear the difference tone of 200 Hz being the modulation detected by non-linearities.

A mathematical examination shows that not only difference tones are produced, but also sum tones. And further there are sum and difference tones between the sum and difference tones! So the whole situation becomes very complex.

Let us now imagine the Helmholtz experiment already described in section 1.15 repeated with pure tones, one of which remains fixed and the other glides slowly upwards. In figure 1.19 the thick lines are the graphs of the two notes. Now any note can produce sum and difference tones with itself and the thin lines represent the sum and difference tones for each note with itself. These are sometimes called 'aural harmonics'.

Suppose the note has a frequency f_1, the thin lines will be at $f_1 + f_1, f_1 + (f_1 + f_1), (f_1 + f_1) + (f_1 + f_1), f_1 - f_1, (f_1 + f_1) - f_1$, etc. In other words $0, 1, 2, 3, 4, 5 \ldots \times f_1$. Mathematicians call a sequence of numbers like 1,2,3,4,5,..., a harmonic sequence and so a series of tones such as $f_1, 2f_1, 3f_1$, etc, in music can be referred to as a set of harmonics on f_1. We shall have more to say about harmonics in sections 2.10, 2.14 and 2.15.

In figure 1.20 we begin to build all the possible combinations such as $f_1 + f_2, (2f_1 + f_2) - f_1, 3f_2, 5f_1$, etc, and in figure 1.21 the same process is taken much further.

Now it can be seen that along the vertical lines where $f_1 = f_2$ and where $f_2 = 2f_1$ the pattern of tones becomes very simple and consists solely of simple multiples of f_1. But at a point where f_2 is marginally higher than f_1, there are literally hundreds of tones.

It is helpful to tilt the book away from you and to view the figure obliquely. Vertical lines can then be seen and it is clear that there are many places where a great simplification in the pattern occurs and the suggestion is that, at those points where the number of different tones present is greatly reduced, the brain will find the resultant sound much more pleasant. The letters at the bottom of figure 1.21 indicate various frequency ratios and correspond to those on figure 1.15 and to the traces of figure 1.16.

A dramatic demonstration of this effect can be performed if two pure

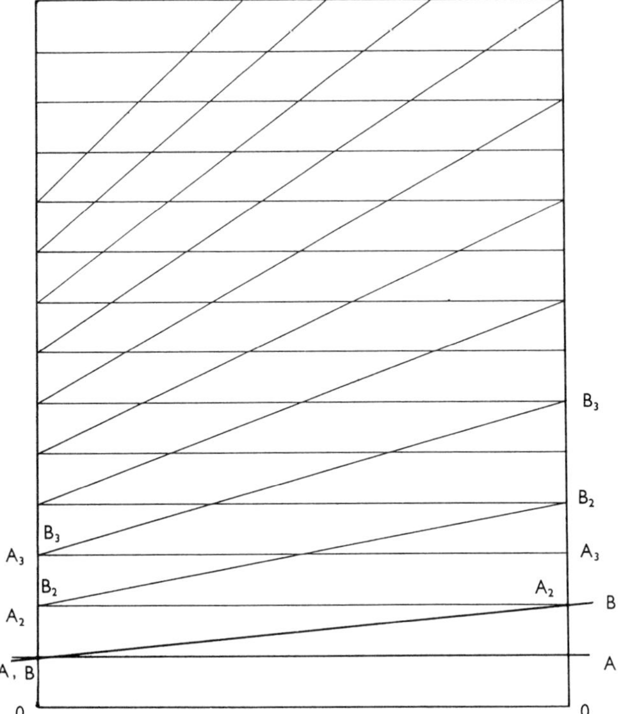

Figure 1.19 Diagram representing some of the results of adding two pure tones. The thick line AA represents a note of fixed frequency. The thick line AB represents a second note which begins in tune with the first note and glides smoothly upwards through one octave. The thin lines represent harmonics and aural harmonics of AA and BB.

tones are recorded on tape, one steady tone corresponding to the tone f_1, and the other a tone gliding upwards corresponding to f_2. The recording is played back at full volume and one can hear quite clearly many different combination tones gliding up and down and corresponding to the various lines with both positive and negative slope in figure 1.20. The demonstration can be enhanced if the tape is re-recorded; the non-linearities are then greater and the effect is even more pronounced.

But that is not the end of the complications; three notes played together such as 400, 480 and 560 Hz should give rise to a difference tone of 80 Hz as the first and probably strongest of many combination tones. And, in fact, one does hear this note quite clearly. The set of three notes is sometimes called a 'tonal complex'. In this particular example the three components are the 5th, 6th, and 7th harmonics of a fundamental of 80 Hz.

The difference tone explanation seems to be satisfactory if the sound

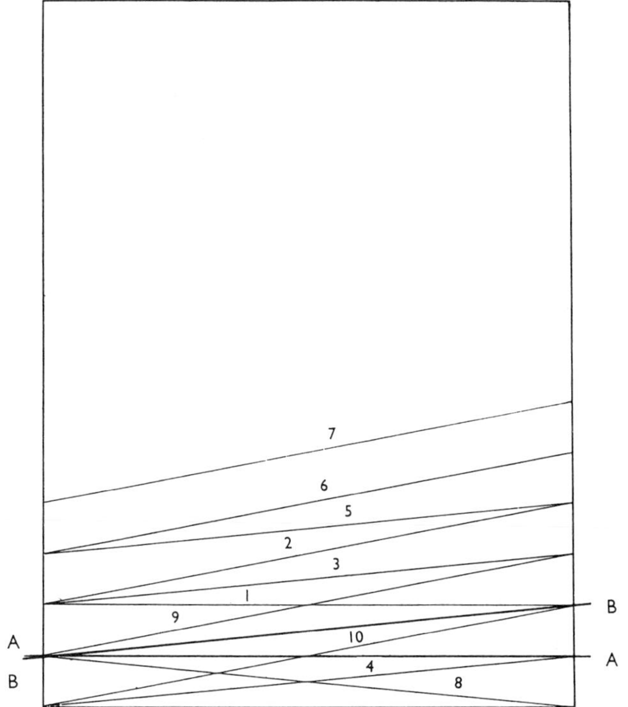

Figure 1.20 Continuation of the development of figure 1.19. The new lines represent some of the combination tones that arise. 1 is 2A; 2 is 2B; 3 is A + B; 4 is B − A; 5 is 2A − B; 6 is 2B + A; 7 is 2B + 2A; 8 is 2A − B; 9 is 2B − A; 10 is 2B − 2A.

is a loud one. But the strange thing is that you can still hear the fundamental note even when the sound is quiet! Now we come to an even more surprising phenomenon. If the three tones are raised by equal amounts to 420, 500 and 580 Hz, and if the origin of the effect were the difference tone, then we should still hear 80 Hz. In fact the note we hear is actually significantly higher! So, obviously, further investigation is needed. This phenomenon is sometimes called the 'residue' effect. We shall come back to this problem again when we consider the tone of the bassoon (section 4.7) in which, although the low numbered harmonics are missing, we hear the fundamental. The effect is sometimes called the 'missing fundamental'.

1.17 PSYCHO–ACOUSTIC COMPLICATIONS

In earlier sections we have mentioned listening tests, for example in connection with the Helmholtz consonance experiment. Such tests play an

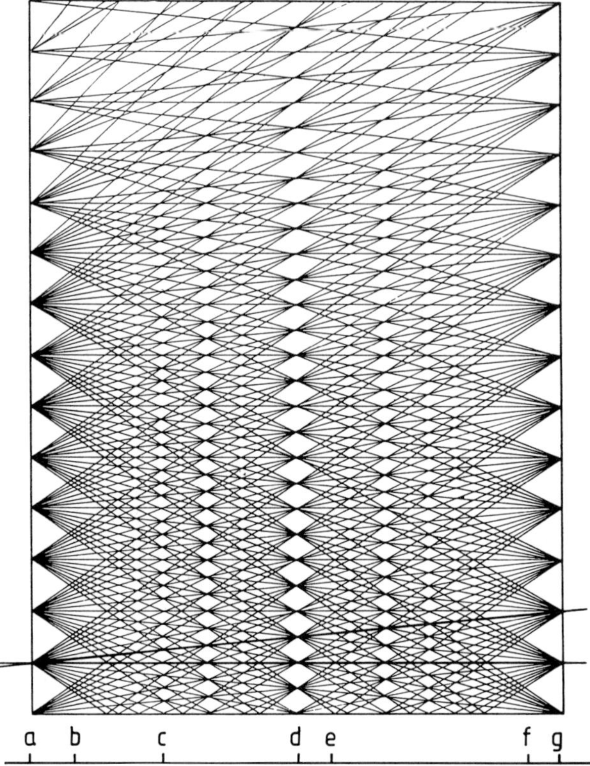

Figure 1.21 Further development of the process begun in figures 1.19 and 1.20. The letters correspond to those of figure 1.16.

enormous part, not only in the specifically scientific aspects of music but also in trying to choose a musical instrument, in trying to place a group of instruments in different quality categories, etc. This therefore seems to be a good point at which to refer to some of the dangers inherent in testing in such a very subjective field. The problem is that, from a purely biological point of view, our brains are continually trying to help us to survive in an unfriendly world. And these attempts on the part of the brain can be a great stumbling block, especially if we are unaware of their existence.

For example, from the earliest moment of our lives our brains are learning. They are storing up all kinds of information in our memory banks and it is essential that we recognise sounds as quickly as possible. Primitive man needed to distinguish, for example, between the sound of a hostile animal moving through the undergrowth and the sound of a harmless animal that might make a good meal. The distinction between the two had to be made in a split second or he might not survive. We still have that facility and it plays a vital part in listening to music. For example, if you listen to an orchestra it is not difficult to concentrate on any one group of instruments, say the strings or the flutes, and hear what they are playing. But that distinction is only possible because the sounds of those instruments are already stored in your memory banks and can be found, compared, and recognised very quickly. If you meet people whose first language is not English you may have difficulty in understanding them at first. But after a day or two you may say to them, 'My word, your English has improved very quickly!' But of course there is probably little change in them; what has happened is that your brain has learned to recognise the sounds they are making. You have been 'programmed' to understand. There are various ways to demonstrate this phenomenon. A recording of speech that has been speeded up without changing the pitch, or has been passed through distorting circuits, may be impossible for an audience to understand on first hearing; but, if they are 'programmed' by telling them what is being said, they find it very difficult to believe that the sentence they now understand is the same one.

And this is exactly what happens in acoustic tests. Suppose you are going to ask a group of people to judge whether certain pairs of notes are 'smooth' or 'rough'. The very act of performing the experiment changes the subject who is doing the experiment. The first time a new pair of tones is heard the brain is trying to identify an unknown sound and, because that particular sound has not been heard before, the response may be somewhat uncertain. The second time that the pair of sounds is heard the brain immediately recognises the sound as one that has been heard before and the response may be quite different.

But the complications and complexities of this area of study are not only scientifically fascinating, but, almost certainly, account in part for the extraordinary aesthetic pleasure that music can afford. If the simple rules held precisely we should probably already live in a world in which all the possible music had already been composed and performed! What a dull world that would be.

1.18 MORE ABOUT THE PART PLAYED BY THE BRAIN

In section 1.11 we discussed some of the problems involved in trying to measure the loudness of sounds and we discussed at length the relationships between quantities that can be measured with a meter, such as sound intensity in dB, and quantities that can only be judged subjectively, such as loudness in sones. Very similar problems arise in connection with frequency. We have already stressed (section 1.7) that frequency is a physically measurable quantity, whereas pitch is a characteristic that can only be judged subjectively. However, whereas in judging loudness the idea of one sound being twice as loud as another has some meaning, though it may be difficult to estimate, the idea of one note having twice the pitch of another does not make much sense. However we do need to spend a little time exploring the complexities of pitch judgment to about the same depth as was done for loudness.

We start by considering the smallest change in measured frequency that we can detect as a change in pitch. We find that over part of the range of hearing a law rather similar to the Weber–Fechner law applies. That is, the smallest perceptible increment in pitch, measured in terms of the corresponding frequency changes increases as the frequency increases. And so, as far as the perception of sounds goes, the smallest musical interval (which of course corresponds to a frequency ratio) that can be detected stays constant, at least over the middle and upper frequency ranges. However, below about 500 Hz the relationship becomes more complicated and it seems that the actual difference in Hertz between two notes that can just be detected stays constant. This in turn means that the musical interval that can just be detected at low frequencies becomes much larger.

Next we need to think briefly about the time it takes for the brain to make a judgment of pitch. Clearly if the time for which a tone is sounded is less than the time required for one complete cycle of the wave there is virtually no evidence on which to base a pitch judgment. Many studies have been made of the whole question of how the brain interprets incoming signals and arrives at a decision on the apparent pitch of the note. The question becomes extremely complicated when the sound being heard is a complex mixture as we saw in sections 1.15 and 1.16

So far we have tacitly assumed that equal intervals judged musically really do correspond to equal frequency ratios throughout the whole range of hearing. It turns out that this is true over the middle frequency range, but at the top and bottom end of the range there are departures from this rule. At the top end of the range the pitch change is usually judged to be less than the simple rule would predict. In other words an interval judged to be an octave musically, may turn out to have a ratio that is higher than 2:1. Similarly at the bottom of the range an interval judged to be an octave may have a frequency ratio that is smaller than 2:1.

Just as we produced a curve showing how loudness is judged subjectively (figure 1.11), we ought to be able to produce a curve showing the relationship between pitch judgments and measured frequency intervals. There is, however no easy way of doing this. As already mentioned the concept of one note having twice the pitch of another does not make sense. In fact when one first considers the problem it seems to be a crazy one. We all seem to judge octaves as a series of recognisable equal steps which turn out to have frequency ratios of 2:1. The solution to the paradox is that these judgments are based on simultaneous comparison. When two notes are sounded together and we judge them to be an octave apart, then the frequency ratios will turn out to be 2:1. And the combined curve for the

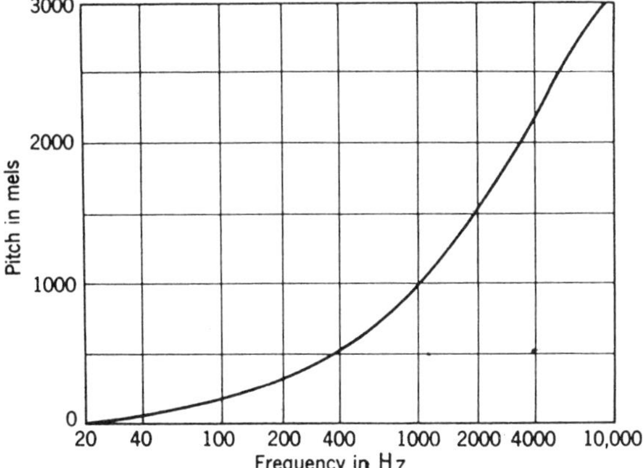

Figure 1.22 Relationship between the subjectively estimated pitch in Mels and the objectively measured frequency in Hertz.

two notes will have the same shape at all frequencies (section 1.15). But if we now attempt to judge the interval by hearing the notes in succession, always using pure tones of a single frequency, and never hearing them simultaneously, the result is different. Musicians would describe the first measurement as being made judging the harmony. The second type of measurement is made melodically and the unit that we introduce to correspond to the sone is called the 'mel'. Figure 1.22 shows the relationship between the mel scale and the frequency scale. However this is an immensely complicated topic with many unresolved controversies and I do not therefore propose to take it further. The moral for physicists is simply that whenever one is trying to make a measurement of a subjective quantity in relation to an objectively measurable one, our marvellous brains take account of so many different variables that it becomes extremely difficult to specify the exact conditions under which the measurement is being made. Without such precise specification we perhaps should not be too surprised that different investigators sometimes seem to find different results.

1.19 CONCLUSIONS

We have discussed a number of very general aspects of music in this first chapter. Some will come up again in later chapters but it seemed necessary to lay down some of the ground rules before proceeding to study the essential features of a musical instrument as distinct from a source of sound.

We may not have answered the question posed in the title of the chapter but we have made a start by asking a number of peripheral questions and seeking answers. Of course these questions have all been slanted towards the scientific aspects of music and it is abundantly clear that a composer, instrument maker, or performer might well have produced some quite different answers.

Answers

The answers to the problem of figure 1.13 are: (*a*) the end of the first movement of the Mendelssohn Violin Concerto, (*b*) a symphony orchestra tuning up, and (*c*) the sound of an audience before a lecture begins.

2

The Essence of an Instrument

2.1 INTRODUCTION

There are many sources of more-or-less musical sounds that occur in every day life; the 'plop' as a cork is pulled out of a bottle, the 'twang' as a stretched rubber band is released, the 'clang' as a metal plate is dropped, or the note produced by running a wet finger round the edge of a wine glass, are all obvious examples. These sounds are musical in the sense that they seem to have a definite musical pitch. But they are not particularly beautiful from an aesthetic point of view. Though even that statement is open to question since the whine of a vacuum cleaner and the note produced by blowing across the neck of a hot water bottle were once used in a concert devised by the late Gerard Hoffnung.

This chapter is concerned, however, with what has to be done to one of these everyday sources of sound to convert it into a usable musical instrument of a more conventional type than those used by Hoffnung. Our concern will be with the method of starting the oscillation in the first place, with the means of feeding in energy to maintain the oscillation, and with the means of extracting the sound energy in sufficient quantities for the sound to be loud enough to be musically useful.

2.2 STARTING A NOTE

We will begin by considering the plop produced when a cork is withdrawn from a bottle. It is clear that as the cork begins to move out, a slight vacuum is caused in the bottle; when it is completely out, air from outside rushes in and slightly compresses the air in the bottle and the sequence repeats with diminishing amplitude as the energy is dispersed by friction between the air and the bottle neck and the viscous drag in the air flow.

Figure 2.1 shows a simple piece of apparatus that can be used to demonstrate the effect. It consists of a plastic tube with a small microphone attached at one end. The microphone is connected to an amplifier and to an

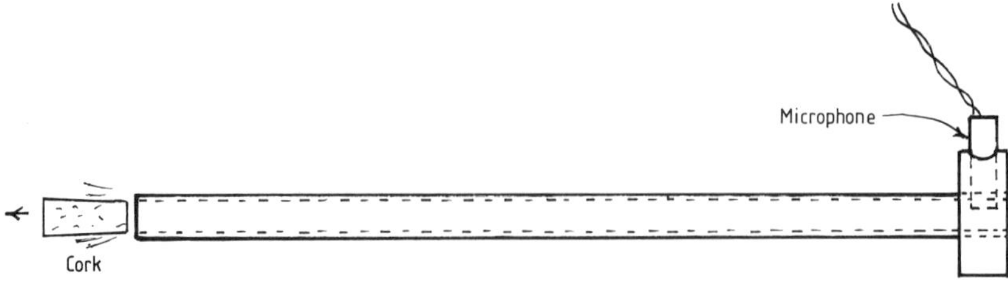

Figure 2.1 Plastic tube with microphone inserted.

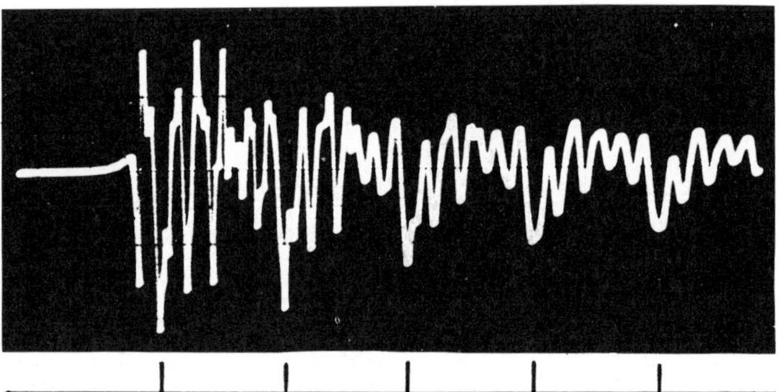

Figure 2.2 Oscilloscope trace produced with the apparatus of figure 2.1 when the cork is withdrawn. The vertical lines at the bottom indicate the natural frequency of the tube.

oscilloscope which is arranged so that a single trace is triggered by the first signal from the microphone. Figure 2.2 shows the trace produced either by withdrawing a cork from the end of the tube farthest away from the microphone, or by striking the flat of the hand on the end of the tube. The oscillatory period of the air rushing backwards and forwards along the tube can clearly be seen and also the gradual decay of the oscillation with time.

There is a Nigerian instrument called a 'shantu' made out of a hollow gourd which works precisely on this principle. The gourd can be anything from about half a metre to a metre long and is five to ten centimetres in diameter with a hole at each end. It is held at about 45° to the vertical in the right hand and the lower end is slapped against the bare left thigh of the player to produce a 'plop' whose pitch depends on the dimensions of the gourd. Large numbers can be used to produce fascinating rhythmic accompaniments (see figure 2.3(*a*)). A helical spring (such as the kind sold as toys which can 'walk' downstairs) provides a model of the behaviour. Figure 2.3(*b*) shows such a spring stretched between the hands of the demonstrator. The sudden inwards movement of one hand sends (*a*) 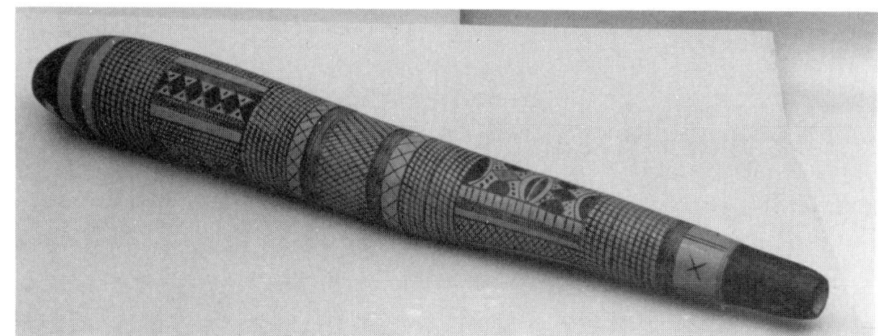

(*b*) 

Figure 2.3 (*a*) A Nigerian 'shantu', and (*b*) a helical spring, designed as a toy, being used as an analogue of compression waves in a tube.

a 'squash' travelling along the spring which is subsequently reflected back from the other end.

All these are examples of longitudinal oscillation; that is to say that the direction of the displacement that occurs is parallel to the long dimension of the object and to the direction of travel of the waves. In section 1.8 we discussed the vibrations of blocks of wood or of pins and of needles. For these the displacements are in directions that are perpendicular to the length of the object, in other words they are transverse.

2.3 NATURAL FREQUENCIES

But in both categories there is a natural frequency that is characteristic of the object. Thus, for the tube or the Nigerian shantu the 'object' is the gas in the tube and the natural frequency depends on the dimensions of the container and on the properties of the gas. (For example, if the tube were filled with helium instead of air, the frequency, and hence the pitch of the note produced, would be markedly different; warm air would give a different pitch from cold air, etc.) For the wooden blocks or pins it is the dimensions and the elastic properties of the material that control the frequency.

The idea of a natural frequency is absolutely fundamental to our understanding of musical instruments. We will start however by considering an example of a natural frequency that has very little to do with music. Galileo is said to have been the first to notice that the time of oscillation of a pendulum is independent of the length of the swing. He is supposed to have watched the heavy oil lamps hanging on the end of chains from the roof of a church. As they swung in a slight breeze he timed the swing using his pulse as a clock and found that whether they swung violently or gently, the period remained constant. It may seem obvious to us now, but, if you think about it, it is very surprising. On the face of it you might well expect it to take much longer for the pendulum to move over a long arc as in figure 2.4(a) than over the short arc of figure 2.4(b). The difference is that, in an arc like that in figure 2.4(a) the pendulum, though travelling further, will travel much faster and so there is compensation. Of course to be absolutely precise the time of oscillation does change a little but, provided the angle of swing does not exceed about five degrees either side of the vertical, the error in assuming that the period is constant for any angle of swing will be less than 1/10 of 1%.

2.4 KEEPING A NOTE GOING

So a child on a swing behaves like a pendulum and has a natural period. But just as with the air in the tube, once started the oscillation will die down; unless we take further action the swing will come to rest. What do we have to do to keep the swing going? Obviously energy has to be fed in and this is most easily done by giving a slight push. But the push must occur at the right moment. If you give a push as in figure 2.4(c), just as the swing reaches the turning point, then you can keep the oscillation going for as long as you like.

But if you give a push at the wrong moment, e.g., as in figure 2.4(d) when the swing is passing its lowest point, disaster is bound to occur. So we now have a way of feeding in energy to keep an oscillation going. A good example of this is in the mechanism of a pendulum clock. Figure 2.5 shows how the shaped teeth of the escapement wheel ensure that the pendulum is given a slight push at just the right moment.

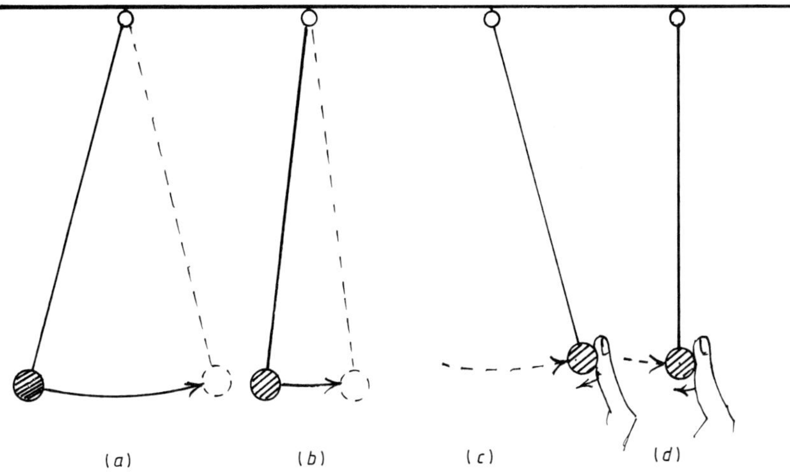

Figure 2.4 A pendulum with (*a*) a long swing, and (*b*) a short swing; with a demonstration of (*c*) a push given as the pendulum reaches the end of its swing which will maintain the oscillation, and (*d*) a push given in the wrong direction at the wrong point of the swing which will stop the oscillation.

While we are thinking about swings and pendulums there is another point to be made that will be useful later. To keep the pendulum, or swing going it is not necessary to give a push at the right moment in *every* cycle. A push in each alternate cycle, or every third swing, or every fourth swing will serve to keep the oscillation going. But, of course, if there are too many swings in between each push, the oscillation will die down a bit and then recover after each push. This turns out to be an important aspect of the behaviour of wind instruments as we shall see in section 4.5. But equally the pendulum can be kept going if a short push occurs at twice, three times, or any other multiple of the natural frequency. Every second, third, etc, push will not connect with the pendulum but there will be one push in each cycle of the pendulum that will connect at the right moment. This point will be taken up again in section 2.10.

2.5 MAKING THE SOUND LOUD ENOUGH: INSTRUMENTS OF THE FIRST FAMILY

The technique of starting an oscillation and then keeping it going by correctly timed pushes is an aspect of the scientific phenomenon called resonance and it is a useful way of increasing the loudness for instruments of family number one. This family was introduced in section 1.9 as the family

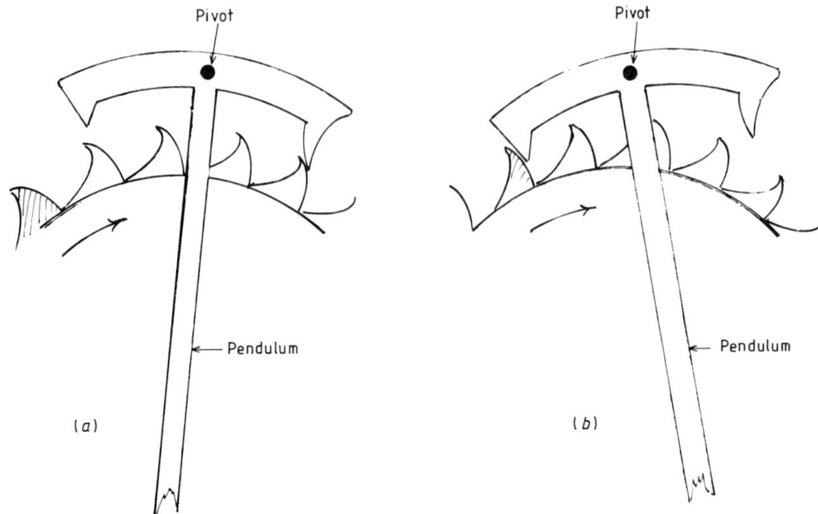

Figure 2.5 Escapement of pendulum clock. The wheel is driven by a falling weight and the teeth are shaped to give the pendulum a push in the right direction once in every cycle. In (*a*) the tooth on the right is about to give the pendulum a push; in (*b*) the shaded tooth has advanced one step.

of instruments that have a separate vibrator for every note. In discussing the piano as a member of this family we stressed that it is possible to adjust each of the separate instruments making up the whole without influencing the rest. This makes possible the use of tuned resonators as a means of making the sound louder (as, for example, in the tubes of a xylophone).

> There are many ways of demonstrating the phenomenon of resonance. One of the easiest methods is to set up a number of pendulums of different lengths, all hanging from a horizontal string as in figure 2.6. One pendulum which has a much heavier bob and is the same length as one of the other lighter pendulums is also hung from the same horizontal string. The heavy pendulum is set swinging and at first all the pendulums move slightly; but very quickly the motion settles down and only the pendulum that is the same length as the heavy one (i.e., has the same natural frequency or period) oscillates strongly.

A very beautiful and striking demonstration of resonance was first described by Dr Higgins in 1777. Michael Faraday investigated the mechanism in 1818, and the apparatus used in demonstrations at the Royal Institution by John Tyndall some 150 years ago still exists. It is called the 'singing flames' and figure 2.7 shows a diagram of the apparatus from Tyndall's

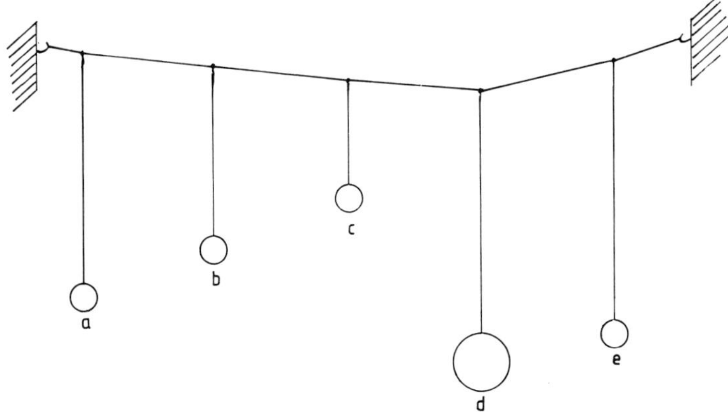

Figure 2.6 Pendulums suspended from a horizontal string. When the heavy pendulum (d) is set swinging, all the pendulums move rather randomly for a short while and then only the one of the same period as (d), i.e., (a), remains swinging for any length of time.

book on sound.

> A very small gas flame is lit at the end of the jet and the glass tube is then lowered over the flame. As the tube is lowered, there comes a point when the shape of the flame suddenly changes and becomes elongated and, at the same time, the air in the tube begins to 'sing'. A relatively pure tone is emitted and it can easily be shown that its pitch corresponds to the pitch produced by blowing across the end of the glass tube. A cardboard tube raised or lowered over the top end of the glass tube can be used to control the pitch.

It seems that a pulse of hot air rises up the tube, is reflected back from the open end as an expansion (see section 4.2) and, on its return, interacts with the flame and so with the gas in the supply tube. The dimensions of the gas supply tube are critical, as is the nature of the gas used. Tyndall's apparatus will not work on natural gas, as we found to our cost many years ago when the Royal Institution was converted from 'town' gas to natural gas. In order to use the apparatus a cylinder of simulated town gas is now used; it must contain about 15% of hydrogen.

If the tube is raised slightly from the singing position and the note is stopped by placing the hand momentarily over the open end of the tube, the note will not start again.

But if the demonstrator then sings the exact note corresponding to that of the tube, the flame will again change shape and the tube will sing. The pitch is fairly critical and, for example if the note is even half a semitone

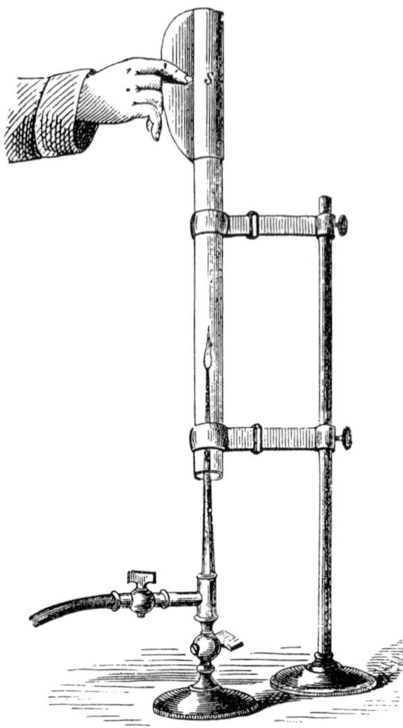

Figure 2.7 The singing flame apparatus as illustrated in John Tyndall's book *Sound*.

flat or sharp the singing will not occur. Obviously the demonstrator's voice excites resonance in the glass tube and this, in turn, excites resonance in the gas supply tube.

Resonance is used in a number of musical instruments, partly to make the sound louder and partly to make it more tuneful. For example the sound made by hitting wooden blocks of a child's xylophone (a direct descendant of the blocks dropped on the floor discussed in section 1.8) is not very loud and not very tuneful. This is largely because the impact between the hammer and the blocks creates quite a large amount of noise and only a relatively small component of pure tone. Resonance in tubes is used to convert it into a much more tuneful musical instrument—the orchestral xylophone. In figure 2.8 the vertical tubes underneath each block can be seen. The tubes are closed at the lower end because the natural frequency of a pipe closed at one end is half that of a pipe open at each end (see section 4.4). This makes it possible to keep the height of the blocks from the floor reasonable for the player. Figure 2.8 is a view looking at the audience side and so the lowest notes are on the right. Why then do the resonating tubes reduce in length as we move from right to left and then lengthen again? In fact this is purely cosmetic. The tubes on the right

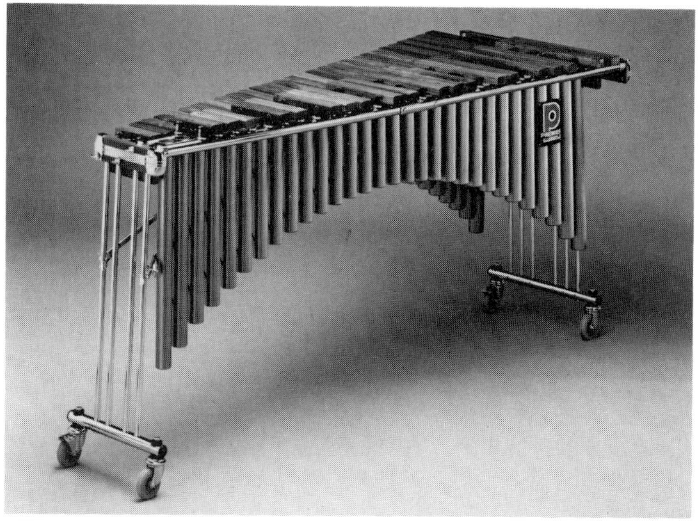

Figure 2.8 An orchestral xylophone. (Photograph by courtesy of Premier Percussion Ltd.)

of the centre are stopped at their lowest end. But the tubes on the left have a stop inserted to create the correct resonating length. The tubes that are visible are those corresponding to the 'black' notes of the scale and the tubes corresponding to the spaces between the notes are dummies and contain no stop.

The effect of the resonant tube can be demonstrated by alternately inserting and withdrawing a hand into the space between the top of the tube and its corresponding block while the block is being struck.

The resonance effect is exploited still more in the vibraphone. The instrument has metal (often aluminium or an alloy of aluminium) plates, whose dimensions are adjusted to give the required notes, and resonators similar to those of the xylophone. But a long spindle runs through all the tubes close to the upper end that can be rotated by means of a motor. Inside each tube the spindle carries a disc which, when vertical, leaves the tube virtually open, but when in the horizontal position nearly closes the tube. When the discs are rotating at an appropriate speed the resonance is varied to give the characteristic vibrato effect which gives the instrument its name. The motor can be turned on and off and, of course, it is important to ensure that the discs are in the vertical position when the motor is turned off.

The resonators used in the xylophone and vibraphone are often described as 'tuned resonators' because they will only operate satisfactorily in increasing the loudness if the natural frequency of the tubes is tuned to be exactly that of the corresponding blocks. Most of the instruments that make use of resonance belong to family number one—those with a separate vibrating device for each note—and also tend to be percussion instruments. The latest addition to this class of instrument is the lithophone being built by Walter Meier and his associates in Zürich in which the tuned blocks are of stone.

Before leaving the subject of tuned resonators there is one more of John Tyndall's classic experiments that I should like to describe. It is in some ways a variant of the 'Wheatstone Telegraphic Concert', already described in section 1.3.

> A very large tuning fork, which is normally mounted on a tuned resonator in the form of a rectangular box, open at one end, is the core of the experiment. The frequency of the fork used in the Christmas lectures was 128 Hz and the inside dimensions of the resonator 160 × 300 × 590 mm.
>
> The fork and resonator are separated and the box is placed on the floor of the theatre with a 30 foot long wooden rod resting on it held vertical by strings. An assistant takes the matching fork and a rubber mallet up into the dome. The whole audience looks up to the dome and sees him strike the fork. He then places the end of the fork on the top end of the wooden rod; the members of the audience immediately drop their eyes to the box at the bottom of the rod because the sound can be heard emerging from the resonator.
>
> Tyndall's manuscript notes of the 1865/6 Christmas Lectures describe the arrangement, ending with the dramatic instruction 'Send Barrett upstairs!'

2.6 MAKING THE SOUND LOUD ENOUGH: INSTRUMENTS OF THE SECOND FAMILY

The second family of musical instruments contains those that have only one, or possibly a small number of vibrating devices, and the pitch is changed by changing the vibrator in some way. For example, in instruments of the woodwind family it is the column of air that is the primary vibrator and the principal way of modifying the frequency is by opening side holes which, effectively, change the length. Instruments like the violin, 'cello and guitar are also members of the second family. The string is the primary vibrator and the pitch is changed by changing the vibrating length of the string. The fact that they have four or more strings is simply necessary to increase the frequency range and to introduce the possibility of playing chords. A

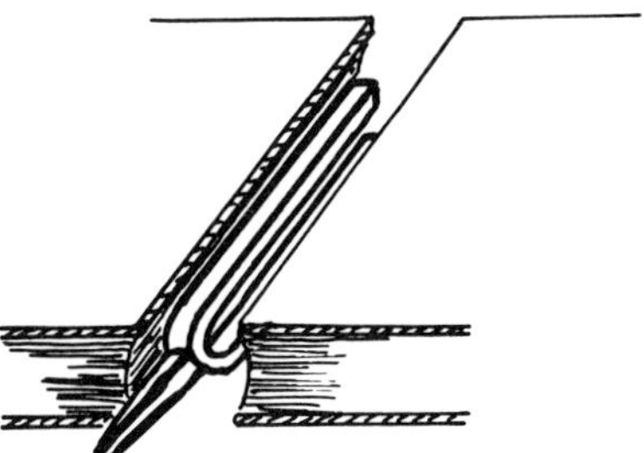

Figure 2.9 Tuning fork placed between two books to increase radiation.

one-string fiddle is perfectly possible and does exist. Clearly instruments that have a small number of primary vibrating devices whose pitch can be changed by the player cannot make use of a tuned resonator since it would have to be tunable to match the frequency of the primary vibrator.

Let us consider the instruments of the string family first. It is quite possible to make a violin with a length of wood, say 5 cm square in section, in place of a body. It can be played but is so quiet that the sound can hardly be heard even by the player. Why? The reason is that the string is so thin that, as it vibrates from side to side the air can move round the string. It is thus unable to create the large pressure differences that are needed to produce a loud sound. The same effect can be demonstrated more easily by means of a tuning fork. The prongs of the fork, like the strings of a violin are small enough for the air to slip round. If a tuning fork is struck and then carefully placed between two books, as in figure 2.9, the sound is noticeably increased in loudness because the air movement is hindered.

If, as in normal use, the fork is struck and then stood on a table or bench, the sound can be heard all over a large room. Now, as the prongs move in and out, the stem moves up and down and the motion is communicated to the large sheet of wood. The air cannot rush round the edges of the table in the time taken for one cycle and hence quite high pressure changes can be produced, consequently leading to a considerable increase in loudness. The question sometimes asked is 'Are we not getting something for nothing from the energy standpoint?' The answer becomes apparent if the time taken for the sound to die away when the fork is on its own is compared with that when in contact with the table. There is clearly no extra energy; but the total energy of the vibrating fork is radiated away much more rapidly and efficiently by the table.

Thus we shall find that improving the efficiency of the radiation of sound from a vibrator is one of the most common ways of increasing the loudness of real instruments.

The bench or table is operating in a different way from that of the tuned resonators of the xylophone. But does the natural frequency of the table or bench not have any influence? A fuller discussion of this is given in the next section (2.7). But we can begin to explore the answer to this question by using the mechanism of a musical box. The essential part looks like a metal comb with teeth of varying length. Pins on the rotating drum pluck the teeth to give the notes of a tune when the handle is turned or the clockwork motor started. If the mechanism is held in the hand the sound is extremely quiet; but if the mechanism is placed firmly on a bench or table top, the sound can be heard all over the room. Now we are increasing the loudness of many different notes and so a tuned resonator cannot be involved. But if the mechanism is placed on a hollow box, or on a metal plate clamped at its centre point (as used for the Chladni plate experiment described in section 3.2) then the quality of the sound is changed; careful listening suggests that notes of different pitches are being amplified by different amounts. So the natural frequency of the object used to improve the radiation efficiency does have an effect on the quality of the sound.

The piano and harp families are examples of instruments that use a sound board or box to improve the radiation. It could be argued that, although we mentioned these instruments as members of the first family, since they have separate vibrators for each note, the sound board, being common to all the vibrators tends to suggest that they have at least one foot in the camp of the second family. The xylophone is, of course, completely in family one. But clearly the design of the board or box is crucial to the sound quality of the finished instrument. Violent resonances must be avoided and the improvement in radiation of the sound must be comparable at all frequencies.

The most elaborate and difficult to make of all sound boxes are the bodies of the instruments of the string family and so we will leave the discussion of these until the next chapter (Chapter 3) which is devoted entirely to that family.

2.7 OTHER CONSEQUENCES OF USING SOUND BOXES

Apart from the desired effect of increasing the loudness of the sounds, there are further consequences of using sound boxes that turn out to be of paramount importance in musical instruments. The first is the effect the sound box has on the way in which the note rises and falls in loudness throughout its duration. The tuning fork is a useful example. The fork is struck and begins to vibrate and emits a very quiet sound. Once it is in contact with a table or a sound box the note becomes very much louder. But what we have to consider now is the transition that takes place just as the fork comes into contact with the table or box. If a fork is struck vigorously and lowered slowly on to the table you can actually hear a 'rattle' just as contact is made. And if a recording is made and the trace viewed on an oscilloscope it can be seen that the transition from low amplitude sine wave to high amplitude sine wave takes several milliseconds. This should not be surprising because, obviously, the table or bench is quite a large object and the whole of it cannot instantly spring into oscillation at the moment of contact by the fork. The table has considerable inertia and all sorts of complex waves radiate along the surface, are reflected back from the edges and so on. But eventually the whole table top begins to vibrate at the frequency of the fork. This transition forms part of what musical scientists refer to as the 'starting transient'. Its great importance, which will crop up many times in the remainder of the book, is that, although lasting for only a small fraction of a second, it is so characteristic of the instrument that it gives the first clue to enable the ear–brain system of the listener to identify it. The main discussion of transients is in sections 2.18–2.20

If the fork is lifted from the table before it has ceased to vibrate, the table will go on vibrating for a fraction of a second and, though the 'terminal transient' is less significant it does play some part in confirming the identity of an instrument.

It may be thought that placing the tuning fork on the bench results in a rather violent transition. But the same principles apply in both plucked and bowed stringed instruments. When the string of a guitar is plucked the energy of the plucked string takes time to be communicated to the body; and when a violin is bowed, not only must the energy of the vibrating string be communicated to the body but the continued vibration of the body at the frequency of the previous note must die away while, at the same time, the body is being driven at the new frequency.

The second major consequence of using a sound box or sound board (already introduced briefly at the end of section 2.6) is that the increase in loudness may not be the same at all frequencies. In other words the natural frequencies of the box or board may be excited into resonance and so the resultant sound owes its harmonic content not only to that of the original vibration, but also to the effect superposed by the sound box or board. The most extreme example of this has already been seen in Tyndall's experiment with the thirty foot rod to the roof in which the box is required to amplify only one frequency—that of the fork.

The effect of the box or board in changing the harmonic content of a sound is sometimes described as a 'formant', or 'formant characteristic'. If you buy a 'Hi-Fi' amplifier (which can be thought of as an electronic version of the sound board) you will be given a graph of the relationship between the degree of amplification and the frequency. Such a graph can be described as a formant characteristic. For the ideal Hi-Fi amplifier most people would consider that a horizontal straight line is the ideal formant—i.e., equal amplification at all frequencies.

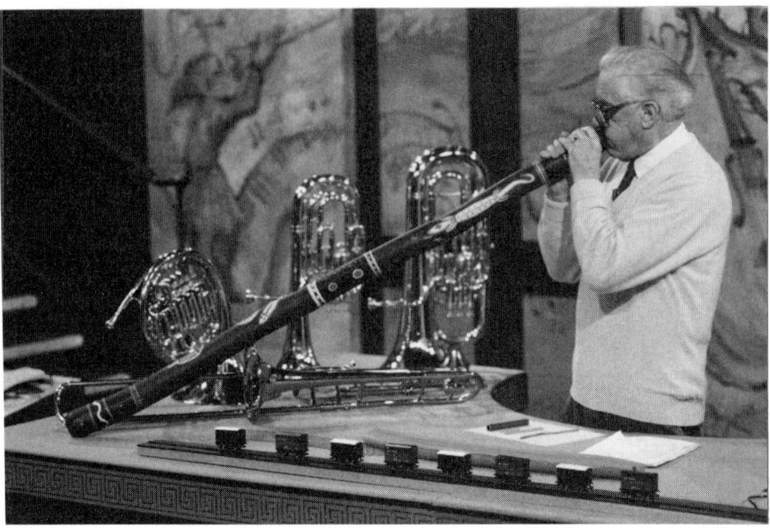

Figure 2.10 An Australian didgeridoo as an example of the maintenance of oscillations at the natural frequency of a tube.

2.8 VIBRATIONS OF AIR IN A TUBE

In section 2.2 we looked at the oscillograph trace of the sound picked up by a microphone inserted in the side and near one end of a tube when a short oscillation was produced by hitting the other end of the tube with the hand. We now need to consider how energy could be fed-in to turn this brief oscillation into a continuous note. We saw in section 2.5 how the natural frequency of a tube can be used as a resonator in an instrument like the xylophone. When the xylophone is being used in performance, pseudo-continuous notes are created by rapid strokes of the hammers on the same block. If, in imagination, we could strike a note on the xylophone with a repetition rate of the hammer equal to the natural frequency of the tube we should create a genuinely continuous note; each transit of the tube by the wave would be slightly reinforced to compensate for losses. This particular exercise is not very practical; but there are many perfectly practical ways of achieving effectively the same thing.

The Australian didgeridoo (figure 2.10) is one example. The lips of the player are made to vibrate, thereby letting puffs of air enter the tube at the right frequency to maintain the oscillation. (It is not too difficult for a brass player to produce a note: the difficult problem is to keep the note going more-or-less indefinitely by circular breathing; that is by breathing in through the nose while breathing out through the mouth.) Figure 2.11 shows a clarinet tube with a pressure microphone inserted in the mouth-

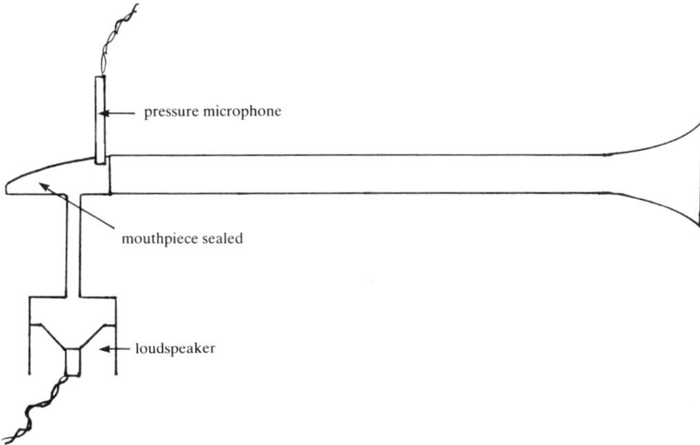

Figure 2.11 Clarinet tube with loudspeaker supplying pulses to maintain the oscillation. . . as done by the player for the didgeridoo. A pressure microphone measures the response.

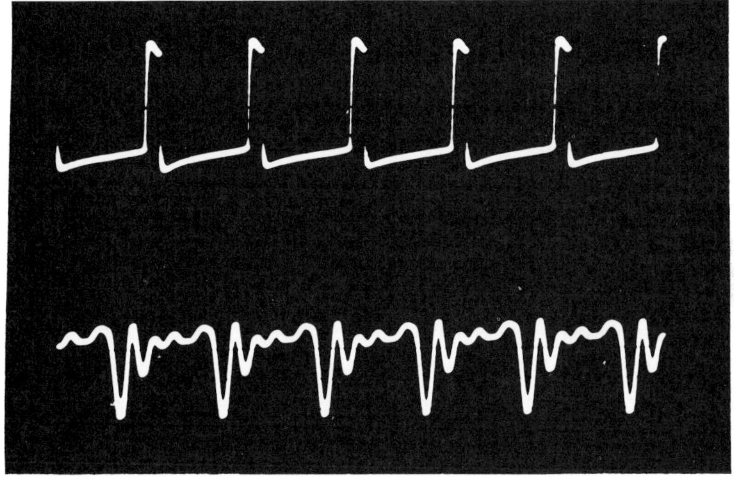

Figure 2.12 Traces obtained from the apparatus of figure 2.11: the pulses are fed-in at the natural frequency (upper trace) and the response (lower trace—which is inverted) is at the same frequency.

piece and a small loudspeaker mounted in a box from which a tube leads to a hole in the other side of the mouthpiece. Pulses can be fed from a pulse generator to the loudspeaker and also to one beam of a double beam oscilloscope. The microphone output is fed to the other beam.

Figure 2.12 shows the kind of result obtained. If the frequency of the pulses matches the resonant frequency of the tube then the pulses (upper trace) have the same frequency as the response (lower trace). Now energy is being fed in at precisely the right moments to keep the oscillation going. Even without the oscilloscope, resonance can be detected by the increased loudness of the response of the tube to the pulses (NB: the lower trace is inverted).

One of John Tyndall's favourite experiments originally performed by Rijke and which is a very dramatic variant of the singing flames can be used to demonstrate the feeding-in of thermal energy at the right frequency to keep an oscillation going. In the version currently used at the Royal Institution a piece of metal tubing 1200 mm long and 60 mm internal diameter (open at both ends) has a crumpled piece of metal gauze (as used to distribute heat from a Bunsen burner) inserted about 90 mm from one end (see figure 2.13). The tube is held vertically, with the gauze nearer the lower end, and is lowered over a large Bunsen burner. When the gauze is incandescent (which can be judged by watching the glow reflected onto the bench round the burner) the tube is removed from the flame. After a few seconds it begins to emit a very loud note. If the tube is turned into a horizontal position the note ceases, but it can be restarted if the tube is returned to the vertical position.

The converse experiment has also been done in which the tube is held with the gauze near the upper end and solid carbon dioxide is dropped in.

This is an excellent demonstration experiment but can hardly be regarded as a means of making the tube into a usable musical instrument. But there are two very practical ways of feeding energy to a tube in order to keep it oscillating that really can make it usable. One is to use an edge tone and the other is to use a reed of some kind.

2.9 EDGE TONES

Probably one of the earliest musical instruments ever made used edge tones. Shepherd boys in the Middle East and elsewhere thousands of years ago used to cut lengths of hollow reed or straw, place a thumb over the lower end and blow across the upper end. Plastic drinking straws (4 mm diameter) make excellent modern equivalents. Lengths of between about 2 and 10 cm will produce penetrating notes which, with a microphone and an oscilloscope can be shown to be fairly pure tones. The frequency, and hence the pitch, depends on the length of the straw. By cutting a series of lengths, sealing one end of each with wax or plasticene, and gluing them to a strip of wood, a playable instrument can be made (figure 2.14). This is a model of the Syrinx or Panpipes.

How is the sound actually produced? Thinking in a very simplistic way you can imagine that the jet of air blown at the back edge of the top of the

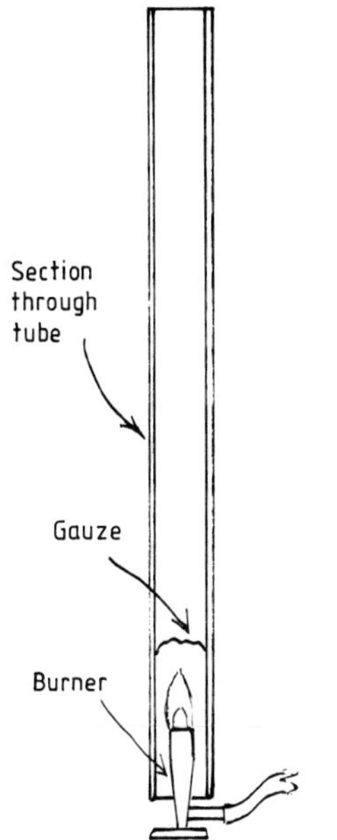

Figure 2.13 Rijke's tube.

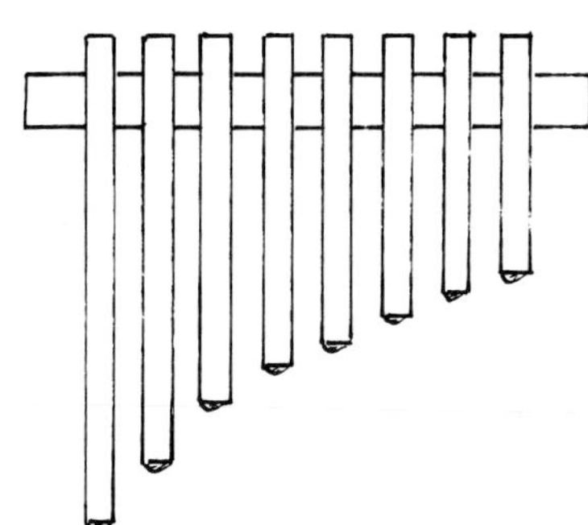

Figure 2.14 Panpipes, or Syrinx.

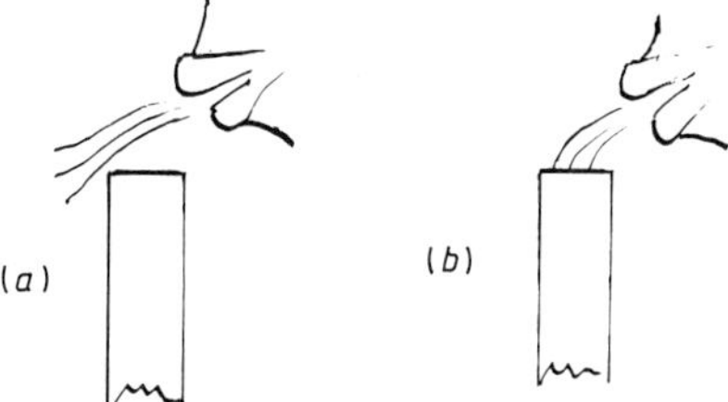

Figure 2.15 Source of oscillations in edge-tone excited pipes.

straw has two choices; it can either pass over the edge or it can turn down into the tube (see figure 2.15). If we imagine that the front of the air jet goes down the tube, then the next part of the air jet will find it easier to go over the top and so on. In other words there is an alternation between the two routes. But the first part of the air jet to go down the tube will be reflected back from the closed end and when it reaches the top again it will react with the air jet. Only if it arrives at the right moment in the cycle will the oscillation continue.

In order that a strong oscillation can be set up there are many factors that have to be just right. The velocity, diameter and exact direction of the air stream must all be right; the stream must be uniform, that is there

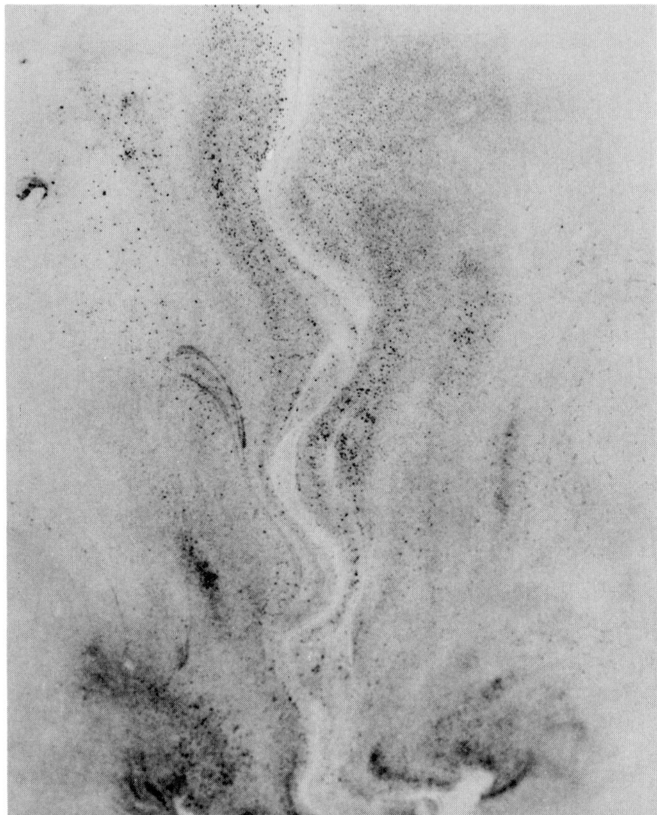

Figure 2.16 Eddies in milk on which charcoal has been scattered.

should be little turbulence within the stream, and the edge of the straw must be clean.

Edge tones or eddies occur widely in many circumstances apart from in musical instruments. A flag on a flagstaff waves because of the eddies produced as the wind blows across the pole. A flag held up by its corners without a pole flutters but does not wave. Edge tones produced as the wind blew down the valley across the Tacoma road bridge in the State of Washington, USA, happened to have the same frequency as the natural frequency of the bridge itself and resonance led to its total collapse in 1940. Figure 2.16 shows eddies produced by moving a vertical stick through a milky liquid made visible by charcoal dust on the surface.

Musically they occur in flue-type organ pipes, and in recorders and flutes. In organ pipes and recorders the streamlined jet is usually rectangular (see figure 2.17) and is directed at a wedge-shaped edge. The thickness of the edge, the angle of the wedge, the angle at which the jet strikes the edge,

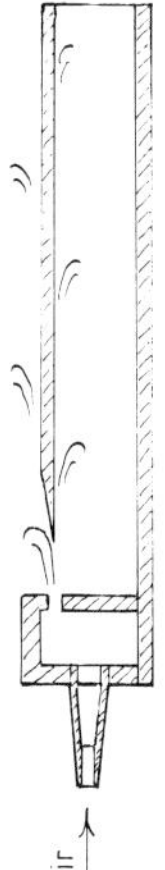

Figure 2.17. Formation of eddies in a simple organ pipe.

and the distance between the end of the slot and the wedge all play an important part in determining the quality and purity of the note, and indeed whether or not a note is produced at all. In the flute and piccolo the variable factors are determined largely by the shape of the aperture formed by the player's lips and the placing of the lips on the instrument.

We shall discuss these instruments in greater detail in Chapter 4 but there is one important question that needs to be answered before we move on. Why is it that if you blow a recorder harder the note may jump to one an octave higher (that is to one at twice the frequency); and if you blow harder still it may jump a further fifth higher (that is to one at three times the original frequency).

2.10 HARMONICS: INSTRUMENTS OF THE THIRD FAMILY

My typical member of the third family of musical instruments is 3.5 m of garden hosepipe. If fitted with the kind of mouthpiece used on a tuba or trombone a whole series of notes can be produced merely by changing the mode or pattern of vibrations inside. We shall discuss this in much more detail in Chapter 4, but for the moment it will suffice to say that the notes produced are referred to as the harmonics of the tube.

What is happening now is that just as the first pulse reaches the opposite end of the tube, another one is sent down and the oscillation can be maintained. The same thing arises at other frequencies. (Compare this with the discussion on the pendulum in section 2.4, in which we said that pushes could occur at higher frequencies than the natural frequency and still keep the pendulum swinging.) These various patterns of excitation of oscillation are called *modes*.

We call the principal note produced by a pipe its fundamental and, since a sequence of numbers like $1, 2, 3, 4, 5, 6, \ldots$, is called by mathematicians a harmonic sequence, we tend to call the notes of frequencies that are whole number multiples of the fundamental, harmonics. Strictly they can be harmonics only if the frequency ratios are *exactly* whole numbers. But in all real mechanical systems, for a variety of reasons, the additional notes are never *quite* whole number ratios of the fundamental. But, for convenience they are often still referred to as harmonics.

There are two other terms that are used in this connection and, since there is often confusion between them it will be as well to explain the differences now. The preferred scientific term for the additional notes is *overtone.* So one might speak of a recorder that is blown so hard that it gives a note that is approximately three times that of its fundamental as producing its second overtone. (The first overtone would be the note of twice the fundamental frequency.) Later on we shall discuss what happens if an instrument oscillates in several modes simultaneously. Then the modes are referred to as *partials* or partial vibrations.

The numbering is always done from the lowest note upwards. The number of a harmonic is *always* the ratio of its frequency to that of the fundamental regardless of whether there are other components present of lower frequency.

The way this works out can be made clear by considering a clarinet which, for reasons that will be discussed later does not produce the second harmonic when it is overblown, but rather the third. That is if you blow harder the note produced is of three times the frequency. This note would be the *first* overtone because it is the first note you come to when you overblow. But it would be the *second* partial, because the fundamental is the first. And it would be the *third* harmonic because its frequency is three times that of the fundamental. Figure 2.18 shows the sequence of harmonics of a particular note, the musical intervals between them and the pitches of the notes involved. It will be immediately obvious that at the bottom end the notes are quite widely spaced and as we move to higher pitches they become much closer together. Thus to play satisfactory tunes in the lower registers it is necessary to fill in the gaps. This is done in the brass instruments such as the trumpet, tuba, trombone, French horn, etc, by changing the effective length of the tube by means of valves or slides. They thus can more properly be described as hybrids, using the technique of family three to make gross changes in pitch and those of family two to make smaller adjustments.

We shall need to explore this still further when we discuss real wind instruments in detail in Chapter 4 and we shall see that all wind instruments make use of the overtones as one of the means of playing tunes or of extending their range. The instruments of family three are, strictly speaking, only the bugle and post horn. But there is one ancient instrument that

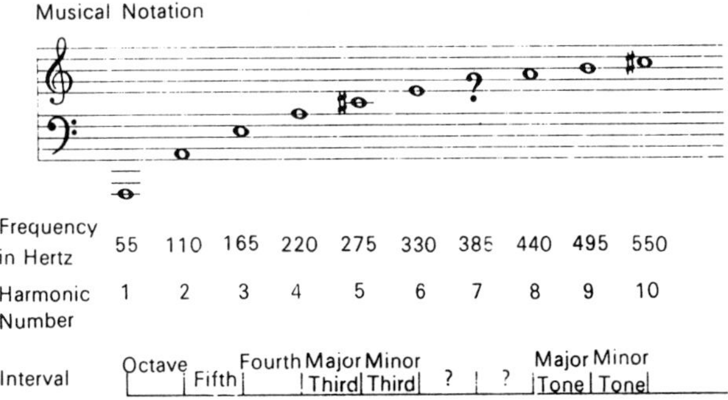

Figure 2.18 Diagram showing the relationship between the harmonic number, pitch of the note, musical notation and the interval between harmonics for a fundamental of 55 Hz.

is worth describing here. It is the tabor pipe. A simple pipe or whistle (like, for example, a tin whistle or flageolet) has a mouthpiece like that of a recorder and has a series of holes that can be covered or uncovered by the fingers in order to vary the effective length of the tube and hence to play tunes. In a tin whistle, to obtain the diatonic scale of 8 notes, the six holes are all covered to give the lowest note. Then the six fingers are raised in turn and the eighth note is produced by closing all the holes and overblowing to produce the octave. But it was common to play the pipe and a small drum called a tabor together and, if the tabor is beaten with the left hand, the maximum number of holes that can be covered while the pipe is also held firmly in the right hand is three (the thumb and little finger are used to hold the pipe). The mouthpiece is designed to aid overblowing and the lowest note used is the octave of the note produced by closing all the holes (that is the second harmonic). The next three notes are produced by successively uncovering the three holes and then the fifth note is produced by covering all the holes again and overblowing to give the octave plus a fifth above the fundamental (i.e., the third harmonic).

2.11 REEDS

At the beginning of section 2.8 we saw that one of the earliest instruments used edge tones as a means of maintaining oscillation. But another possible mechanism is the reed, and we saw how oscillation could be maintained by the lips of the player acting as a reed in a didgeridoo. Many instruments use a man-made reed and it is possible to make a working model using a 4 mm plastic drinking straw. A piece about 10 or 15 cm long is a good starting point. At one end the straw is flattened by drawing a blunt blade along it for a length of about 3 cm (see figure 2.19(*b*)). The end is then

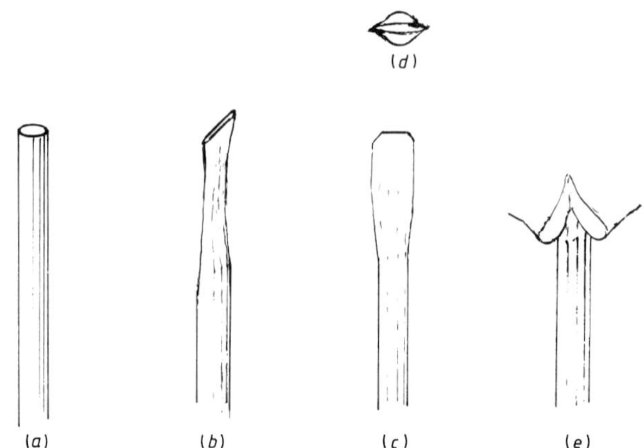

Figure 2.19 Making a reed pipe from a 4 mm diameter plastic drinking straw. (*a*) the straw; (*b*) the end of the straw flattened; (*c*) the end trimmed square and the corners cut off; (*d*) end view of flattened straw; (*e*) the whole of the flattened portion must be inside the mouth.

cut square and the corners are cut off as shown in figure 2.19(*c*). The end view of the reed should now look something like figure 2.19(*d*). To play the reed, it is placed entirely within the mouth with the lips closed round the straw below the reed (figure 2.19(*e*)). Blowing sharply down the pipe should now produce a more or less musical squeak. Tunes can be played by cutting holes in the side at the lower end of the straw. The oscillograph trace for a note on this instrument is nothing like a pure tone and is much more like a succession of pulses like those produced with a bagpipe chanter (see figure 2.34).

This very simple device is a working model of the class of instruments known as 'wind cap' instruments. Their key characteristic is that the lips of the player do not touch the reed. Figure 2.20 shows two examples of wind-cap instruments: (*a*) is a bagpipe practice chanter and (*b*) is a close up with the cap removed to show the reed; (*c*) is a modern version of a cornemuse and (*d*) is a close up with the wind cap removed to show the reed. The chanter, or melody pipe, of the bagpipes is also, of course, a wind-cap instrument. In order to explore the action of the reed we made an oblique cut across the wind chamber of a practice chanter and cemented it on a sheet of glass as shown in figure 2.21. A hole is also bored in the side with a flexible tube attached so that the chanter can still be blown. A television camera can now view the end of the reed; the oblique glass plate avoids troublesome reflections. Illumination is provided through a fibre-optic cable from a lamp and between the lamp and the end of the cable is a rotating disc of holes whose speed of rotation is variable. By adjusting the wind pressure and/or the speed of rotation, the stroboscopic

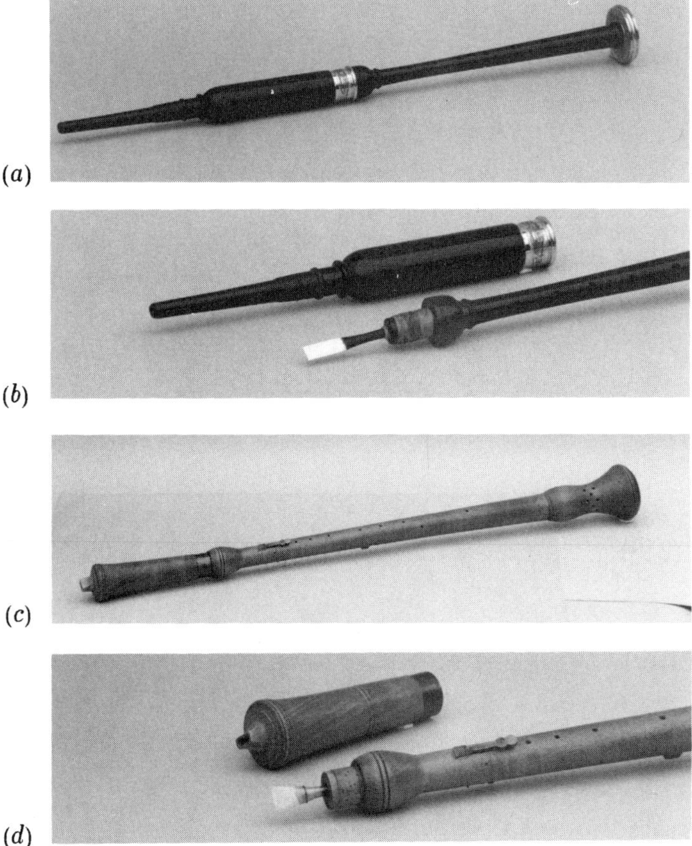

Figure 2.20 Instruments of the wind-cap type: (*a*) bagpipe practice chanter; (*b*) close-up of chanter reed; (*c*) cornemuse (modern); (*d*) close-up of cornemuse reed.

effect can be used to see the reed slowly opening and closing.

With edge-tone excitation of a pipe the pressure changes are relatively smooth and by blowing gently it is possible to excite almost exclusively the fundamental frequency of the pipe. But with reed excitation it is clear that a much more complex mixture is produced.

2.12 ANALYSING MUSICAL SOUNDS

The next question concerns how we can investigate such mixtures to find out if there really are different harmonics present. Before the days of electronics the device most often used was the Helmholtz resonator. Figure 2.22 shows a set that belongs to the Royal Institution. Each spherical shell has two openings. One is a tapered tube that can be fitted into the ear, and the

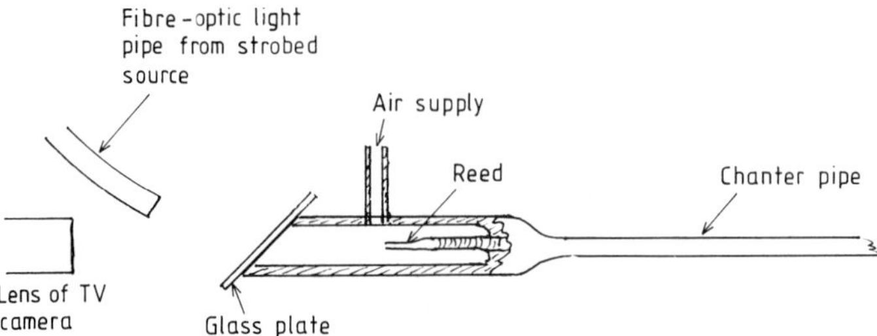

Figure 2.21 Modification of a practice chanter for stroboscopic viewing of the reed.

Figure 2.22 A set of Helmholtz resonators from the Royal Institution.

other is a short cylindrical tube that is to be pointed towards the source of sound being considered. Helmholtz developed a theory that links the total volume of the sphere and cylindrical tube, the area of cross section of the cylindrical tube, and the natural resonant frequency. (It may be of interest to know that the frequency is directly proportional to the fourth root of the area of the tube, and inversely proportional to the square root of the volume.) The resonance is quite sharp and if, for example, the nozzle of a resonator is placed in the ear and then a gliding sine tone is played there is no difficulty in identifying the point on the glide at which resonance occurs. Sets of resonators were produced, each engraved with its resonant frequency. There also exist tunable resonators in which the cylindrical tube is longer and has another tube sliding in it so that the total volume can be varied. A set of lines is engraved on the inner tube so that the outer tube can be set to given frequencies within a small range.

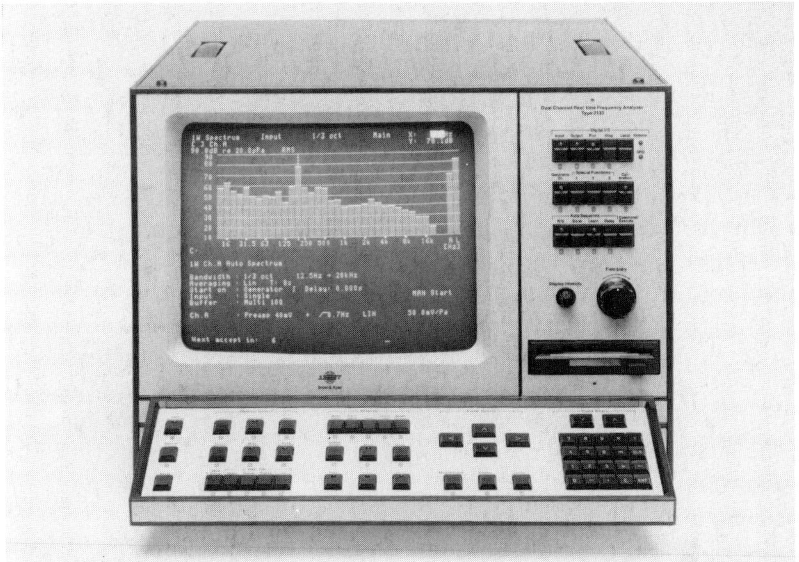

Figure 2.23 A real-time digital frequency analyser. (Photograph by courtesy of the makers, Brüel & Kjaer Ltd.)

One of the important uses of these resonators was to identify the separate harmonics or partials present in a complex mixture of tones.

During the hundred years or so that have elapsed since resonators were first used a great many different methods of frequency analysis have been used. The fundamental principle of all of them is really the same, except that instead of mechanical resonators, electronic systems that have resonant frequencies are used. Perhaps the most useful of these are the type known as 'real-time frequency analysers'. The most recent ones convert the sound to be analysed into digital form (see section 5.8) and then perform the analysis using computer-type programmes. (An example of this kind was lent by Brüel and Kjaer during the lectures and is shown in figure 2.23.) The result of the analysis is shown as a bar chart or histogram which indicates the relative proportions of the various frequencies present more or less instantly, and so the changes with time can readily be seen. (This latter feature is the origin of the adjective 'real-time'.) Figure 2.24 shows some examples of different wave forms, together with the corresponding analyses to illustrate how the system operates.

2.13 WHY DO REEDS PRODUCE SO MANY HARMONICS?

It is easy to show that reed instruments of all types tend to produce richer mixtures of harmonics than flute-type instruments using one or other of the frequency analysing techniques that we have mentioned. In fact it is hardly necessary to do the analysis; you can actually hear the components if you listen carefully. But why should reed excitation produce so many more harmonics?

The following demonstration may be helpful in making the reasons clear. In figure 2.25(A) a series of light pendulums of differing lengths is suspended from a horizontal rod which itself is suspended. A much heavier pendulum is arranged so that it can either drive the horizontal rod continuously, or it can give the rod a short push once in every cycle. When the heavy pendulum is connected so that it drives the horizontal rod continuously, all the pendulums start to swing for a brief time but very soon the system settles down and only the light pendulum whose period is the same as that of the heavy pendulum remains swinging. In other words excitation with a sine wave (such as is produced by an edge tone) tends to produce only the fundamental frequency and no harmonics.

This is not too surprising if you think it through carefully. Consider a light pendulum whose frequency is twice that of the heavy pendulum. For part of the cycle the two will be in opposition and so any oscillation started is rapidly damped out. The same analysis can be done for any other pendulum except the one that is in tune.

But now consider what happens if the linkage is arranged to give a short push once in every cycle of the heavy pendulum (figure 2.25(B)). Now we have the situation that we have already discussed in section 2.4. It is not necessary to give a push once in *each* cycle; once in every other cycle, or every third cycle will do equally well to keep the oscillation going. So any pendulum whose natural frequency is a multiple of that of the heavy pendulum is set in motion.

To complete the picture, if we disconnect the heavy pendulum and just give the horizontal bar a single push then all the pendulums start swinging regardless of their length. The acoustic parallel for this is that a single pulse or click will excite a tube of *any* length. The tube with the microphone insert described in section 2.2 is an example. Another example would be any of the percussion instruments such as bells, gongs, cymbals, etc, where a single blow can produce complex mixtures of harmonically unrelated tones.

2.14 HOW WE PERCEIVE MIXTURES OF HARMONICS

Before we move on to consider other consequences of the way in which notes

Figure 2.24 *Opposite.* Examples of the use of an analyser such as that of figure 2.23: (*a*) pure tone; (*b*) analysis of (*a*). (*c*) Saw-tooth wave; (*d*) analysis of (*c*). (*e*) Train of pulses; (*f*) analysis of (*e*).

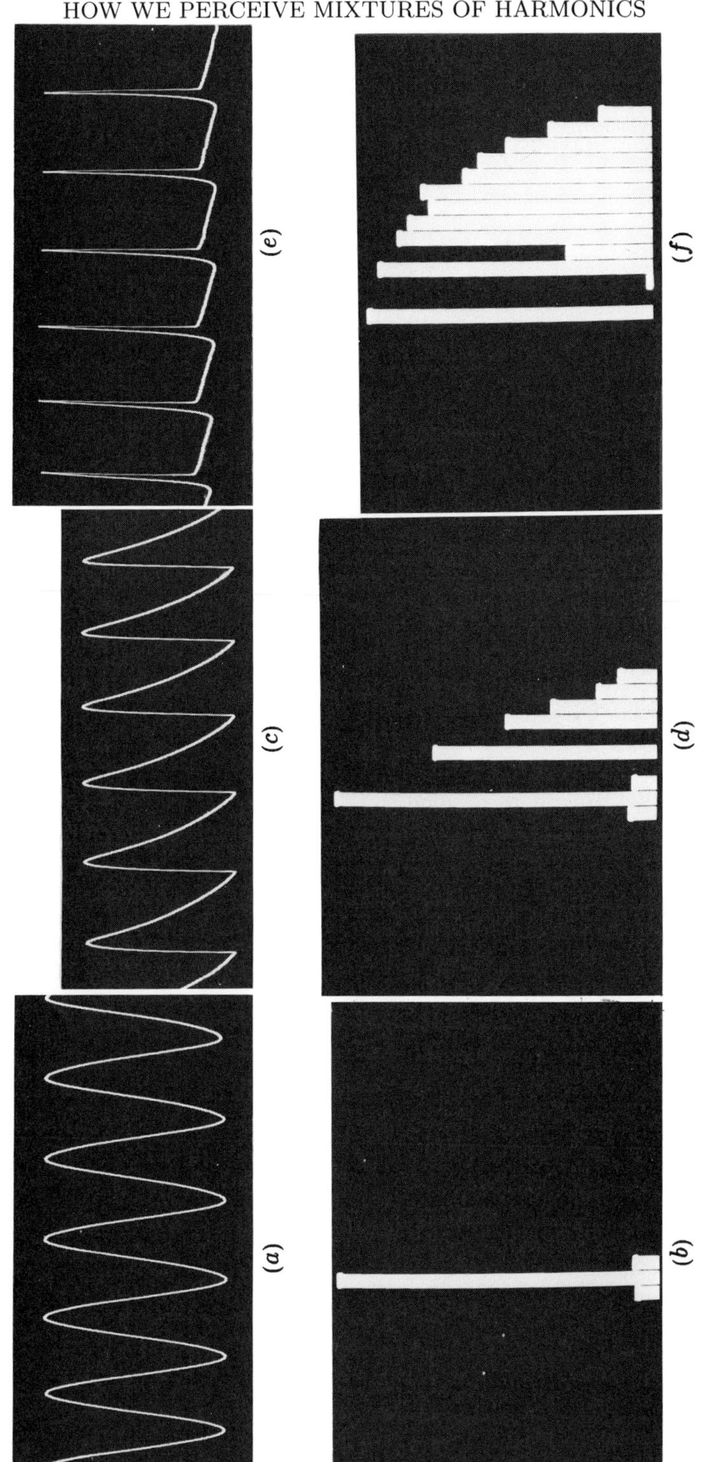
(a)
(b)
(c)
(d)
(e)
(f)

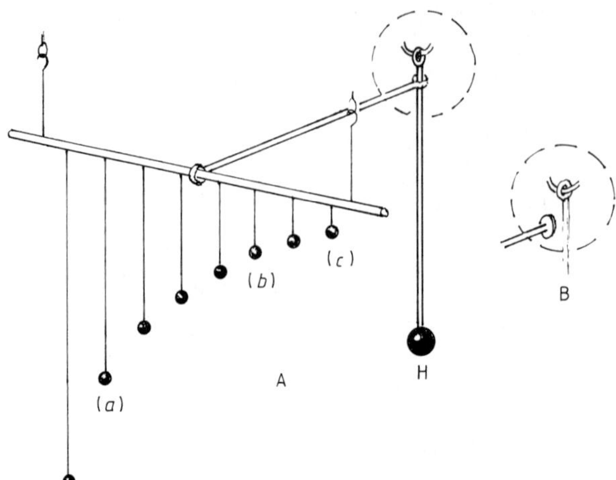

Figure 2.25 Set of pendulums that can be driven by the heavy pendulum H. One of the pendulums (a) has the same natural frequency as H; two of the others (b, c) have frequencies that are harmonics of that of H. The rest have frequencies not directly related to that of H. When the linkage is in place for the whole time (A) only pendulum (a) responds. When the linkage is disconnected to give a short push once in each cycle, (B), (b) & (c) respond. When one single push is given all the pendulums respond.

commence it will be worth our while to discuss the way in which our ear–brain system responds to mixtures of harmonics. An electronic keyboard capable of producing pure tones corresponding to exact multiples of a given fundamental either separately or simultaneously provides a useful way of investigating the problem.

Start by playing a single pure tone that can be regarded as a fundamental of frequency f. Then, while keeping that sounding, add a tone of twice the frequency, $2f$. With a little practice you can easily hear the two notes separately even when they are sounding together. But, if the two notes are initiated simultaneously, it is then much more difficult to separate out the notes.

The effect becomes even more striking if the first five notes of a harmonic sequence ($f, 2f, 3f, 4f, 5f$) are sounded consecutively and each is kept sounding once it has started. Within the resultant complex tone you can pick out each separate harmonic. But now if that complex is stopped and all five notes are restarted simultaneously the result is perceived as a complex tone and it is much more difficult to persuade your brain to listen to the separate notes.

If different combinations of harmonics are tried (e.g., $f + 3f + 5f$, or

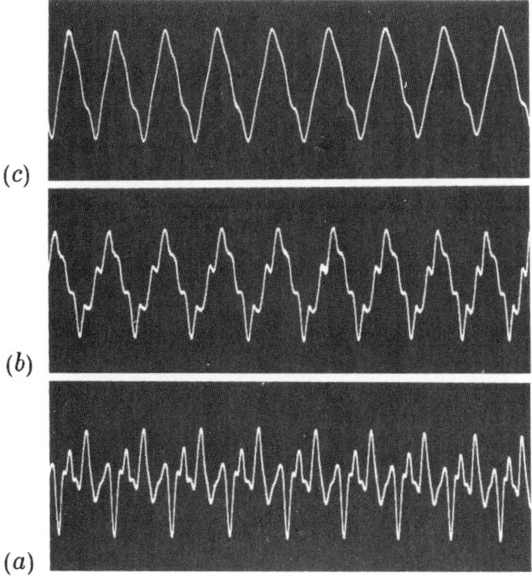

Figure 2.26 Trace for three clarinet notes an octave apart from each other: (*a*) is F_3 in the Chalumeau register (174.5 Hz) recorded and played back normally; (*b*) is F_4 in the Clarino register (349 Hz) recorded normally but played back at half speed so that it sounds to be at the same pitch as (*a*); and (*c*) is F_5 in the high register (698 Hz) recorded normally but played back at a quarter of the speed so that it sounds to be at the same pitch as (*a*).

$3f+4f+5f$) the resultant compound note will sound significantly different in quality but it will always be easier to disentangle the separate components mentally if they are brought-in in sequence than if they are initiated simultaneously. If the amplitude of each component can be changed then even greater variations in the quality of the resultant sound can be produced. The late Professor Arthur Benade introduced the term 'recipe' for the particular combination of partial components in a particular tone. The real-time frequency analyser can be used to show the difference in recipes not only for different instruments, but also for different notes on the same instrument. The difference in recipe for different notes on the same instrument is brought out very clearly by making a recording of three notes an octave apart on the clarinet. The highest note is then played back at one-quarter speed, the middle one at half speed, and the lowest one at the correct speed. All three notes now sound at the same pitch, but the differences in quality are immediately apparent, both to the ear, and to the eye as in figure 2.26 which shows the oscillograph traces. We shall discuss some of the consequences of these effects in greater detail in Chapters 3 and 4.

2.15 HARMONICS OF STRINGS

The following description of a useful way of demonstrating harmonics on strings is taken from my book *The Art and Science of Lecture Demonstration.*

> There have been many suggestions for the best kind of cord with which to demonstrate harmonics on strings. Some people use gas-tubing, some ordinary clothes line, I have even met the use of rubber tubing filled with mercury! But I have no doubt that the best material is 5 mm diameter soft white rubber cord of the type that used to be used for vacuum seals. A length of about 5 m seems to be the most useful. One end is held firmly by an assistant and it is important that the assistant does not attempt to move in sympathy with the demonstrator at the other end. I send a wave along the rope by plucking it as though it is a large scale model of a guitar string. This has two purposes; the explicit one of showing the audience that when a string is plucked a wave actually travels up and down, and a more covert purpose which is to give me an idea of the fundamental frequency of the cord and to adjust the length and tension to get this in the right range. Then, having got the fundamental frequency in mind, I can oscillate my end of the rope up and down at that frequency and hence produce the fundamental mode with a node at each end without too much trouble. But it took quite a lot of practice to be able to do this with confidence. And it is important from the psychological point of view to make it look easy...otherwise the audience are not so ready to accept this mode of vibration as something that is a natural to the string rather than as some trick of the demonstrator.
>
> Once this has been done it is relatively easy to obtain other modes by oscillating at approximately twice, three times, etc, the fundamental frequency and it is usually possible to obtain at least the fifth mode. But it is rather like playing a musical instrument: it is no use trying to force the rope; you have to be sensitive to its natural frequencies and adjust very rapidly until the desired one is found, when quite a large amplitude can be maintained with very little movement of the hand holding the end. There is no substitute for lengthy practice!

There is, in fact, a connection between the initial demonstration of plucking the string and the later demonstrations of successive modes. It can be shown mathematically that the pattern obtained by plucking is the sum of a sequence of separate harmonic modes. The theorem known as Fourier's

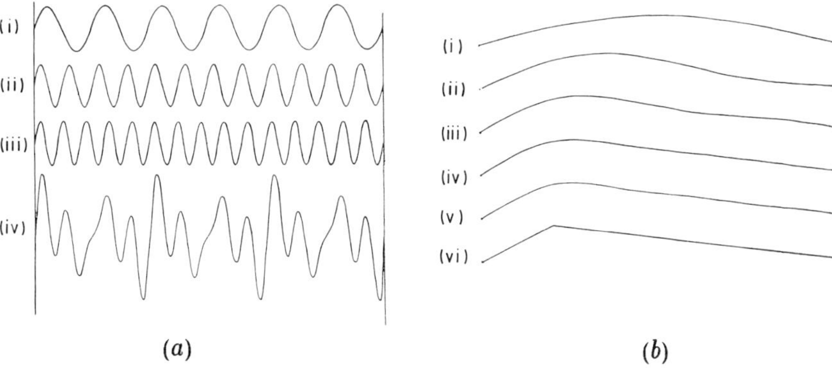

Figure 2.27 (*a*) Calculated curves of three pure tones at (i) 6 Hz, (ii) 12 Hz, (iii) 15 Hz, and (iv) of their sum. (*b*), (vi) is a drawing of a plucked string and (i) to (v) are the successive summations of the first six harmonics. (Since the amplitude of the fifth harmonic is zero there are only five summations shown.)

theorem suggests that any waveform that repeats periodically (say at frequency f) can be represented as the sum of a harmonic set of sine waves. Although it is difficult to prove mathematically, it can be shown experimentally that a particular waveform corresponds to a unique set of amplitudes and relative phases of harmonics. Figure 2.27(*a*) shows an example of a summation performed mathematically over a time period of one second. The three components are of 6 Hz, 12 Hz, and 15 Hz. In other words they are the second, third, and fifth harmonics of a 3 Hz wave. The calculated sum of the waves repeats at 3 Hz (i.e., the fundamental) although no component of that frequency is present (see section 4.7)

Figure 2.27(*b*) shows how the reverse process works. The wave shape produced by displacing a string of length 1 m by 10 cm at a point 20 cm from one end (vi) can be analysed mathematically to give the first six harmonic components with amplitudes as follows: 7.5 cm, 3.03 cm, 1.32 cm, 0.47 cm, 0, and −0.2 cm. (The negative sign indicates that this component is out of phase with the others.) The remaining five curves show the effect of successive addition of the components. Note that the amplitude of the fifth harmonic is zero, corresponding to the fact that the string was plucked at a point one fifth of its length from one end.

In practice the modes of a string are not *exactly* harmonic. The amount by which their frequencies differ from whole numbers depends on the stiffness of the string, on the rigidity of the end supports and on various other factors. In a piano, for example, where the strings are under very high tension and are very stiff, a partial such as the 25th might have a frequency as much as 20% different from its true harmonic value.

John Tyndall used a number of demonstrations to show the modes of a stretched string. He was particularly concerned that the whole audience should be able to see clearly what was happening and this presented some difficulties in the semicircular lecture theatre at the Royal Institution. If the stretched string or rope were placed along the front edge of the lecture bench, the members of the audience sitting at each side would have difficulty in seeing. One solution was to mount the rope hanging vertically down from the roof and for the lecturer to set the vibrations going by shaking the lower end. One slight disadvantage of this was that the weight of the rope itself meant that the tension at the lower end was less than that at the upper end. But then this could be turned to advantage as a demonstration of the fact that the velocity varies with the tension. Since the frequency is a constant factor the wavelength varies along the rope. But there is still the problem that the vibrations are in a particular plane and hence the section of the audience whose line of sight was along this plane would not be able to see. He therefore connected the lower end of the rope to a vertical spindle which, by an arrangement of pulleys could be made to rotate. By varying the speed of rotation the different modes could be excited and the whole audience could see the nodes with rotating sections of rope in between.

> A most elegant demonstration that has been repeated many times is also due originally to Tyndall. In the modern version a length of resistance wire such as Nichrome is attached at one end to an electromagnetic vibrator that can be fed from an audio oscillator of variable frequency. The other end passes over a pulley and a weight is attached. The tension is adjusted until a convenient set of vibrational modes can be set up by altering the driving frequency. By means of two crocodile clips, attached at each end of the wire, a high current can be sent through it, so causing it to heat up. By careful adjustment of the heating current it can be arranged that the wire is on the point of glowing when it is not vibrating. As soon as it starts to vibrate in a particular mode the wire that is moving rapidly from side to side between the nodes cools, its resistance drops and so the total current through the wire increases and the nodal points where the wire is not moving begin to glow brightly. In a darkened room the nodal points stand out very clearly.

Harmonics can be demonstrated on real instruments such as a violin or 'cello by lightly touching the string at points a half, a third, a quarter, etc, of the length of the string from one end instead of pressing the string down to the finger board as in normal playing. These tones are used as special effects in music. Figure 2.28 shows how lightly touching the string at a particular point eliminates all but one mode.

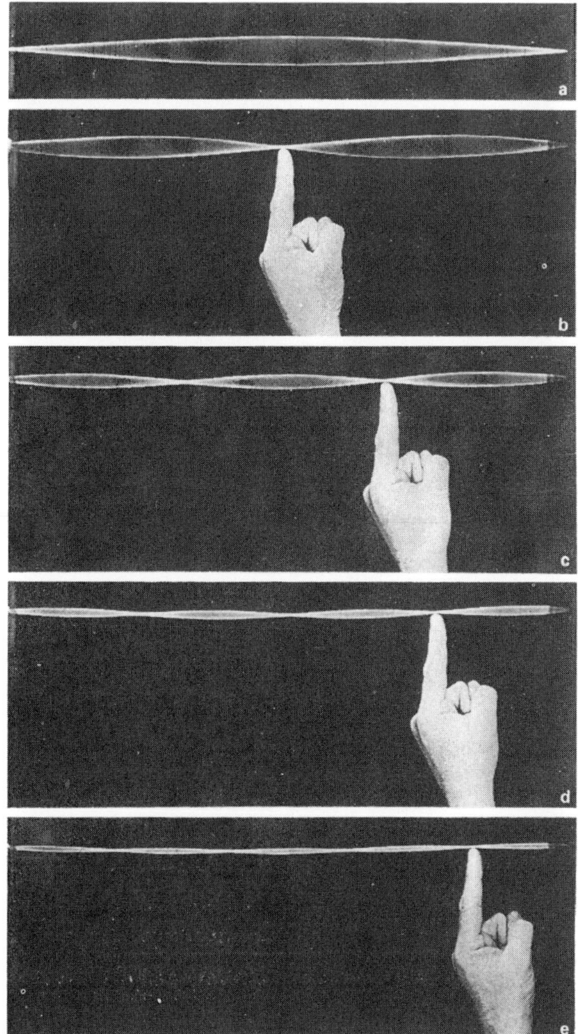

Figure 2.28 Vibrating rubber cord stopped with a finger at various points to produce successive harmonics.

2.16 KEEPING STRING VIBRATIONS GOING

In section 2.8 we discussed ways of keeping the oscillation in a pipe going. Now we must consider how energy can be fed in to keep a string vibrating. The commonest method that is widely used in real instruments is bowing. The physics of bowing is quite fascinating and very complicated. It depends primarily on the peculiar frictional properties of resin which, in turn, lead

(*a*)

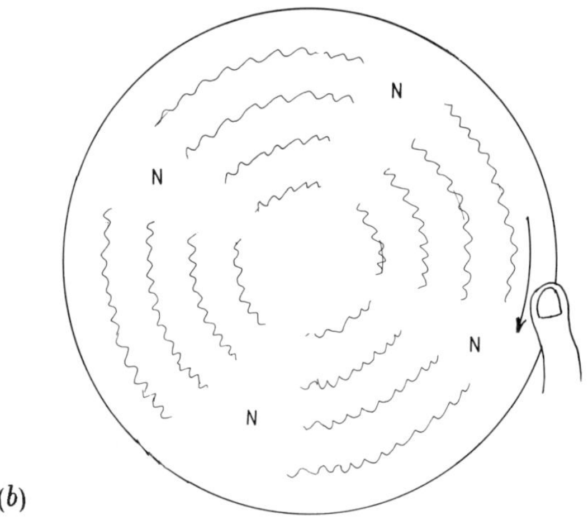

(*b*)

Figure 2.29 (*a*) Excitation of Wheatsone's 'big bowl' by rubbing the rim with wet fingers. (*b*) Drawing of the pattern of ripples produced, showing the four nodes (N) which rotate with the finger.

to a kind of motion known as 'stick–slip' motion.

Perhaps the most familiar example of this kind of motion experienced by non-musicians is the effect that gives rise to the expression 'squeaky-clean'. Glass or china that is very slightly greasy gives no noise when stroked with a finger. But if it is completely grease free it will emit a squeak. The after-dinner trick of running a wet finger round the rim of a wine glass to set it into vibration uses this property.

Wheatstone had a large bowl made to demonstrate the effect and Tyndall used to show how the movement of the water surface as the rim was stroked revealed how the rim was divided into four vibrating sections with still points or nodes between each (see figure 2.29). The frequency of vibration can be changed by partially filling the glass with water; the more water, the lower the frequency because the mass of the water is, in effect, added to that of the glass. In an open glass like this the volume of air plays little part in the oscillation (but refer to section 3.3 in which the relative parts played by the air and the solid container for various shapes of vibrator is discussed).

An easy way to demonstrate stick–slip motion that is more closely related to the bowed stringed instruments is to use cotton gloves coated with powdered resin and lengths of 5 mm solid brass rod. Lengths between about 300 and 1000 mm are most useful. A rod is held firmly at its midpoint in the left gloved hand and the rod is stroked from the centre outward with the right hand. It is essential that the rod is completely grease free and I usually wipe the rod with a little alcohol and then rub it briskly with a resined glove before doing the experiment. The forefinger and thumb of the right hand are used and, once the rod starts to emit a note, only a very light grip is necessary. What happens is that the resin has high static friction and low dynamic friction. When the right finger and thumb grip the rod, the high static friction allows the rod to be very slightly stretched between the two hands. But the restoring forces set up in the rod soon rise to a level which overcomes the sticking forces and the fingers slip. Once the fingers start to move relative to the rod, the friction drops to the low dynamic value and a considerable movement occurs. The fingers then stick again and the cycle is repeated. Each time the fingers slip a pulse travels to the end of the rod and back and so a compression is communicated to the air. The length of the rod and the velocity of the pulses along the rod determine the frequency at which pulses are emitted and hence the frequency of the note. The rod has been set into longitudinal vibration and will continue to emit a note as long as the stroking continues. A shorter rod will obviously give a higher pitched note as it takes less time for the pulse to traverse the rod. A series of rods, each clamped at its midpoint can constitute a kind of musical instrument (see figure 2.30). The reason why the rod is held or clamped at the midpoint is to ensure that the mode of vibration is the one in which the midpoint is a point of no displacement and each end moves symmetrically in and out.

In the bowed strings the bow itself carries a collection of hairs, which are usually from a horse's tail. They are covered with tiny scales which enable the powdered resin to stick to the bow. When the bow is placed on the string it sticks. Then, when the bow starts to move laterally the

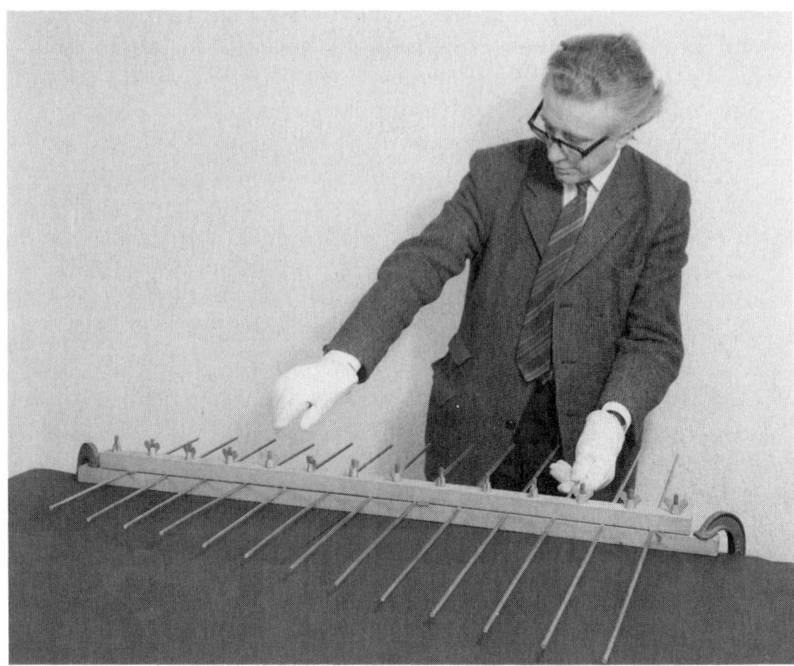

Figure 2.30 'Rodophone' consisting of brass rods stroked with resined gloves.

string is drawn to one side until the component of the tension in the string at right angles to the string is high enough to overcome the static friction. Dynamic friction takes over and the string slips back to its other extreme position. On its return there will be a moment when its velocity is the same as that of the bow and static friction will momentarily take over and the whole cycle repeats. So the lateral motion of the bow across the string feeds-in oscillatory energy to keep the transverse vibrations of the string going.

It is interesting to note in passing that, in the extraordinarily thorough investigations into vibrations of plates made by Michael Faraday in 1831 (see section 3.2), one of his experiments involved a parchment sheet, tightly stretched across the mouth of a funnel, to the centre of which was attached a single horse hair.

> Upon fixing the funnel in an upright position, and after applying a little powdered resin to the thumbs and forefingers, drawing them upward over the horsehair, the membrane was thrown into vibration with more or less force at pleasure.

The great Indian physicist Sir C V Raman made a study of bowing as long ago as 1918 using a laboratory machine that could make a bow travel

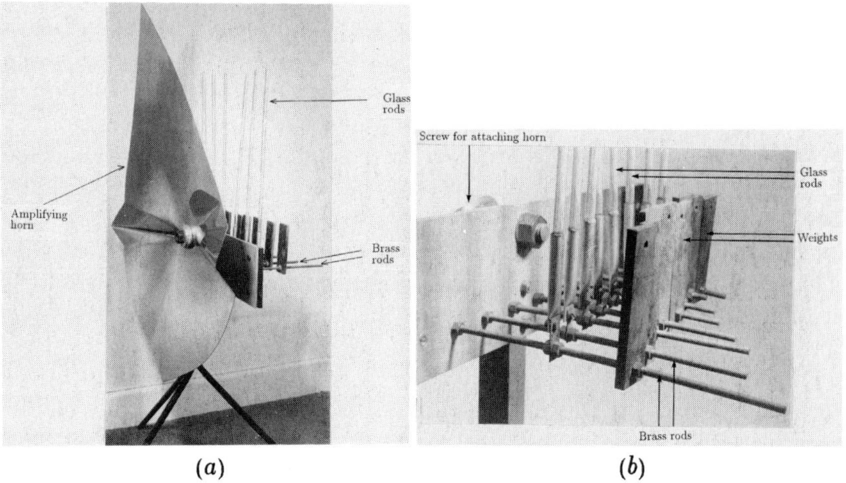

Figure 2.31 (*a*) A complete 'Structure Sonore', and (*b*) close up of the mechanism of (*a*).

across the strings of a violin. Many later studies have been made and most recently computer models of the behaviour of bowed strings have given considerable enlightenment. But the behaviour of a bow on a string in the laboratory is very different from that of the bow in the hands of a skilled violinist in the middle of a concert. We shall return to this point in section 3.8 in our study of real stringed instruments.

2.17 A CONTEMPORARY MECHANICAL INSTRUMENT

We will now look at a contemporary instrument that beautifully illustrates the essential features of any instrument and will serve to give a resumé of what we have said so far. The instrument is one of a family designed by a French group called 'Les Structures Sonores Lasry-Baschet'. I first met François Baschet in the late 1960s and was immediately impressed with the way in which, though designed for music, his instrument provided a beautiful model of the scientific essentials of a musical instrument. He gave me permission to make a copy of one of his smaller instruments (sometimes called a 'cristal') and the illustrations of figure 2.31 are of my model.

We start with a separate vibrator for each note. In this case they are lengths of screwed brass rod, R1, R2, R3, etc, each firmly fastened to a slab of steel, S, which links together the vibrations of each separate rod. In order to increase the inertia of each rod, and so to prolong the vibration, each carries a block of steel, B1, B2, B3, etc, whose position along the screwed rod can be varied for tuning purposes. Then we need a means of feeding-in energy to start and maintain the vibrations of the rods. This is provided by a series of solid glass rods, G1, G2, G3, etc, connected to the screwed rods via phosphor bronze springs. The glass rods are about 5 mm in diameter and about 30 cm long. The position of the springs on the rods is very critical and the adjustment of the positions of the blocks B and the glass rods G are to some extent interdependent and the tuning process is one of trial and error.

The instrument is played by stroking the glass rods with wet fingers. The glass rods go into longitudinal vibrations (as did the brass rods in section 2.16) and this in turn sets up transverse vibrations of the screwed brass rods.

But the instrument is very quiet at this stage. We need some means of increasing the effectiveness of the radiation into the air and this is achieved by the addition of a stainless steel 'horn', H. The increase in volume when the horn is added is almost unbelievable, and, once the horn is in position, a single finger-tip run lightly up and down a well-wetted glass rod will produce a note that can be sustained almost indefinitely.

Of course the precise shape of the horn, H, not only alters the volume of sound, but also its quality, in exactly the way we saw with the musical box in section 2.6.

2.18 HOW NOTES CHANGE WITH TIME

The horn of the instrument described in the last section also has an effect on the way in which the note changes with time. Once the horn is attached it takes a longer time for the note to build up to its loudest level and the quality actually seems to change during the note build-up. The way a note changes with time, in fact, turns out to be one of the most important characteristics in enabling our ear–brain system to recognise the instrument. In order to explore this feature it will probably be easiest to consider the way in which attempts were made, starting in the 1930s, to synthesise the sounds of real instruments by electronic means. We shall have much more to say about the technical details in Chapter 5, but we will consider the general principles here.

In section 2.14 we talked about the recipe for a musical sound, in other words, ways in which the particular mixture of partial frequencies or harmonics influences the tone quality of the resultant sound. Physicists in the 1930s were well aware of the appearance of the oscillograph traces of the wave forms generated by various instruments. Figure 2.32(a) shows the kind of trace that was studied in that period and you will find such traces in many textbooks published between about 1920 and 1950 which would be labelled:

(i) wave trace for a flute;
(ii) wave trace for a clarinet;
(iii) wave trace for a guitar.

To begin with these are very incomplete descriptions of the origin of the traces. For example (i) should be labelled 'Wave trace for a one-hundredth of a second sample of a steady note, (say A_4) played on a particular flute, at a particular loudness by a particular player, with a particular lip formation, in particular acoustic surroundings.' Any variation in any of these designations would change the trace. But for our present purpose the most significant one is the time duration. In one one-hundredth of a second there is virtually no variation. Once the note has begun the same shape is repeated regularly at the fundamental frequency corresponding to the note. Fourier's Theorem (already introduced in section 2.15) states that if a function is periodic at a certain frequency then it can always be represented as the sum of an infinite series of harmonics of that frequency with particular amplitudes and phases. Naturally the scientists of the day assumed that Fourier's theorem would apply here and the earliest synthesizers (called electronic organs at that time) simply provided a means of generating such a series of harmonics and of controlling their relative amplitudes to change the waveform, and hence the tonal quality.

But organs like the Hammond and the Compton, both developed round about 1932 (see section 5.6), while producing pleasant and useful sounds could not equal the sounds of real organ pipes. Listeners could immediately recognise the sound as 'electronic'.

If we look at traces of the same notes and instruments as in figure 2.32(a) but lasting for longer periods—one tenth of a second for 2.32(b) and a whole second for 2.32(c)—we immediately become aware of two of the factors that distinguish these sounds from electronically generated ones. An electronic pure tone remains precisely periodic no matter what the time duration; the real instruments show marked departures from exact periodicity for the longer durations. Also the real instruments in figure 2.32(c) show a slow build up during the first fraction of a second whereas an electronic note can start almost instantly.

The ear–brain system of a listener is very quick to recognise both these features. This is not too surprising. Consider if you were walking out in the country and you heard a steady, high-pitched whine. You would be unlikely to ask 'Who is singing that note?' or 'I wonder what animal that is?' You would be much more likely to ask 'What *machine* is that?'

The brain seems to recognise the slight irregularities present, especially in a voice or the sound of a wind instrument. We shall consider why there are irregularities when we look at the voice in section 4.21. The most characteristic is the slight wobble called vibrato, the frequency of which is usually about 7 Hz. It is very easy to imitate vibrato electronically, but the

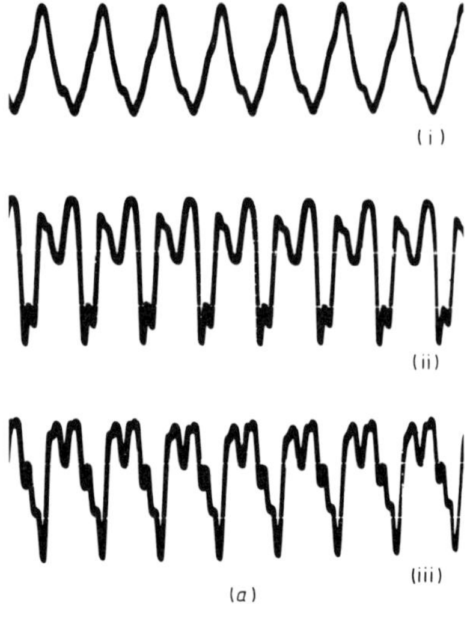

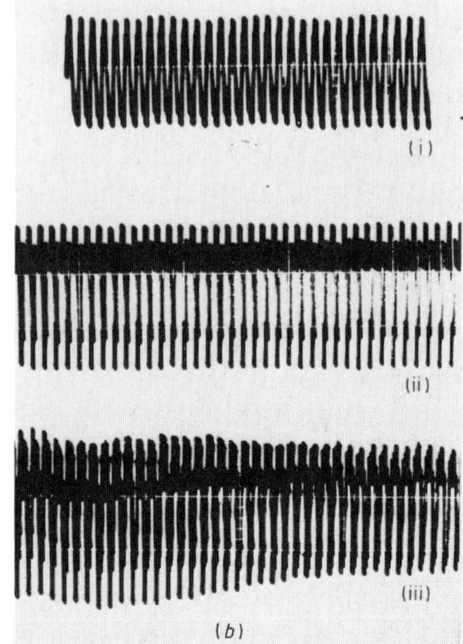

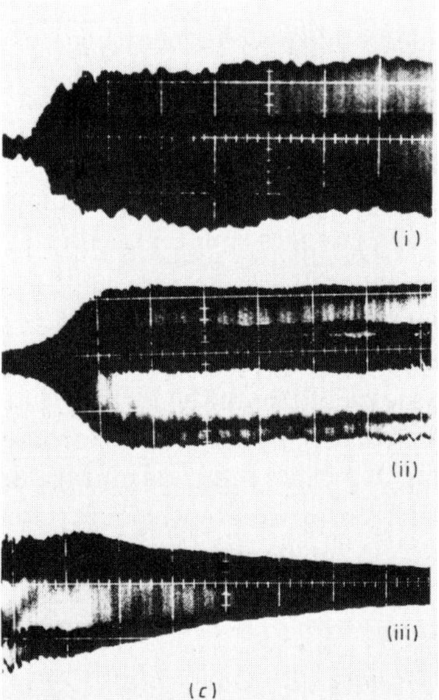

Figure 2.32 (*a*) Traces of notes lasting 1/100 second. (i) is for a flute, (ii) is for a clarinet and (iii) is for a guitar. (*b*) Traces of the same three notes again, but lasting 1/10 second. (*c*) Traces of the same three notes yet again but lasting 1 second. The scan is now so slow that the individual oscillations cannot be seen and the record is virtually that of the envelope of the wave.

notes still sound electronic because the brain recognises that the wobbles of the vibrato are themselves too regular. Indeed the brain is extraordinarily adept at spotting the complete regularity that is the feature of synthetically produced sound. Only in the last ten or so years has it become possible to imitate the irregularity of voices or instruments sufficiently well to deceive the ear–brain system.

The overall way in which the note changes in amplitude with time is often called the 'envelope'. Indeed the trace in figure 2.32(c) is so slow that the individual oscillations cannot be seen and these three figures are virtually the envelopes for the waves. In some of the early (1960s) true electronic synthesizers (as opposed to electronic organs) simple envelope shapes could be imposed at will and produced surprising results. For example if a simple saw tooth waveform, (like, for example, that of figure 2.24(c)) has superimposed on it a triangular envelope in which the note commences suddenly and then decays to zero over an interval of half a second or so, it begins to sound like a plucked guitar string. If, on the other hand, a symmetrical triangular envelope is used so that the note rises and falls slowly, the sound is somewhat reminiscent of a bowed violin.

2.19 THE ALL-IMPORTANT BEGINNING OF A NOTE

Although the overall envelope is extremely important in aiding recognition of an instrument, one part turns out to be much more significant than the rest. That is the very beginning of the envelope—the so-called 'attack' or 'starting transient'. There are many different ways of demonstrating the significance of the beginning of each note in the process of instrument recognition by the brain. The simplest, in that it involves no special electronic 'tricks', involves at least three instrumentalists, for example clarinet, flute and French horn, or any other combination. They are asked to play facing away from the audience or from behind a screen. At a given signal one begins to play a steady note, a few seconds later another instrument begins to play a second note that harmonises with the first, and finally the third instrument begins to complete a chord. The audience has no difficulty in deciding the order in which the three instruments began to play. Now the experiment is repeated, but this time all three begin to play at the same time and then drop out in turn at intervals. Now it is far more difficult to judge the order.

The reason is that, in the first experiment, the ear–brain system recognises the transient of each instrument as its note comes in. In the second experiment the transients are heard at the beginning, but the only change when an instrument drops out is to the combined harmonic structure, which is much more difficult to recognise than the transient.

A tape recorder that records the full width of the track at once, and hence will play a tape backwards if the reels are reversed provides a very interesting source of demonstrations. A recording of a tune played from the last note backwards is made on any solo instrument. When this tape is, in turn, played backwards on the tape recorder, the tune comes out the right way round, but the envelope shape of each note is reversed.

The piano is a particularly good subject for this demonstration since not just the transient, but the whole envelope shape of each note is very recognisable. When the tape is played backwards the instrument sounds at first like some kind of organ, or harmonium. This is because the initiation of each note is now gradual, rather like that of an organ, and the harmonic mixture is similar to that of an organ. After a very short while, however, the listener begins to reassess the original diagnosis and has doubts. This is because the piano initiation is very sudden and so, when reversed, the note rises in loudness and then comes to a sudden stop—which is not characteristic of any particular instrument. Thus, after preliminary identification as an organ, the brain is unable to confirm and so there is confusion.

A third way to demonstrate the importance of the transient is to record a long note on a tape recorder and then to wipe out the beginning, including the transient, so that the note commences very suddenly. The effect is most striking and it is difficult, for example to distinguish between a trombone and a French horn, or between muted trumpet and an oboe.

In section 5.6 we shall discuss both envelopes and transients again in relation to the development of electronic synthesizers, and there is one final point to be made here. It is that the implication so far has been that both the envelope and the transient are overall amplitude effects. In fact in recent years it has been shown that it is necessary to consider the change in shape of each separate component of the recipe of a note. In other words there are changes of timbre with time during the transient period and in the remainder of the note.

Figure 2.33 illustrates this point. The frequency analysis of the note C_4 of a small harpsichord is shown in figure 2.33(a) and its waveform in figure 2.33(d). This sound was recorded almost immediately after the note was struck. The sound for figures 2.33(b) and (e) was recorded 1/4 second after the note was struck, and the sound for figures 2.33(c) and (f) was recorded 3/4 second after the note was struck. The different rates of decay of the frequency components and the change in waveform with time can both be seen clearly.

Figure 2.33 *Opposite.* Illustration of the differing decay rates for the note middle C (263 Hz) on a small harpsichord. (a), (b) and (c) represent the frequency analysis, with (d) the wave trace immediately after the peak amplitude has been reached, (e) the wave trace about a quarter of a second after (a) and (d), and (f) the wave trace a further half a second after (b) and (e).

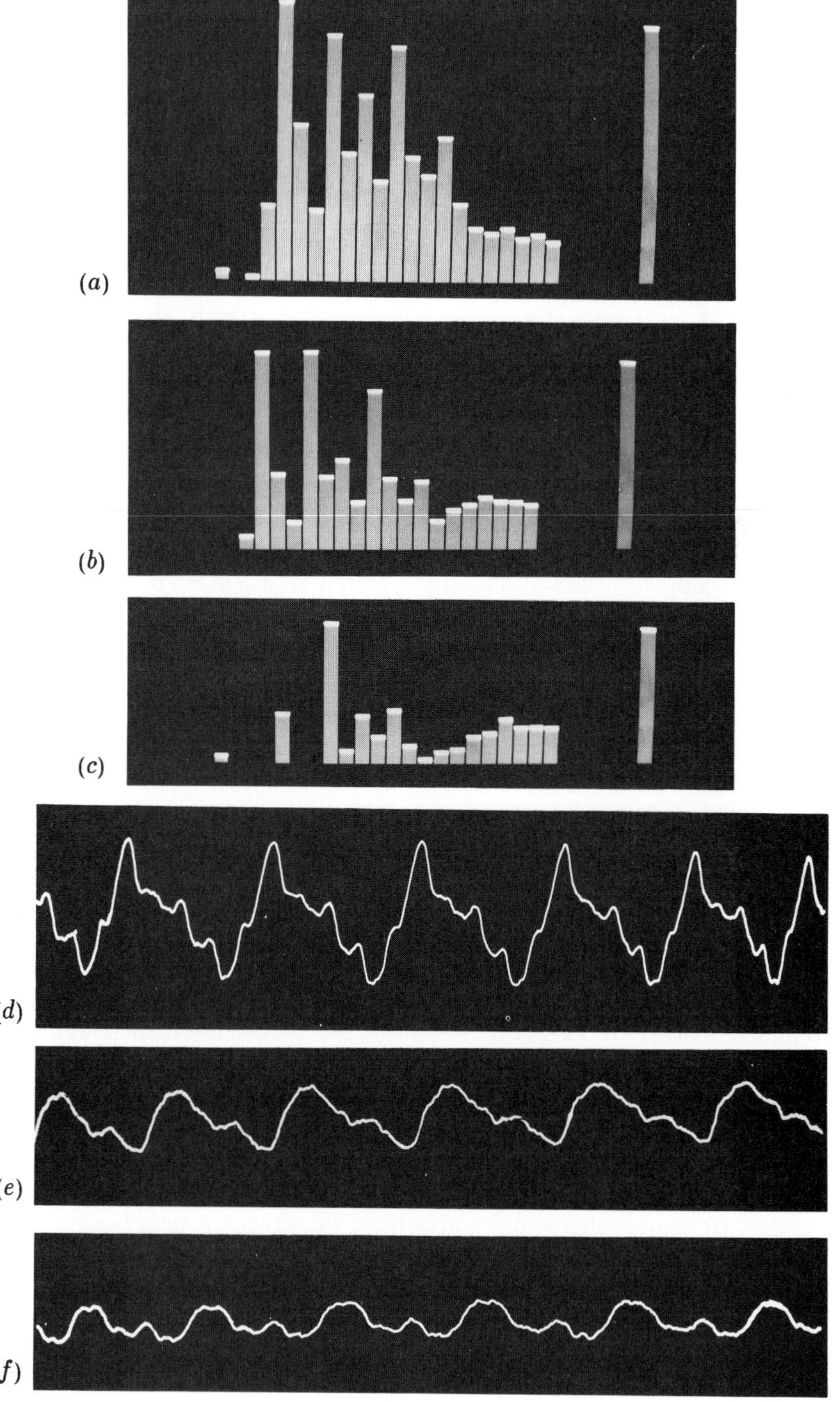
(a)
(b)
(c)
(d)
(e)
(f)

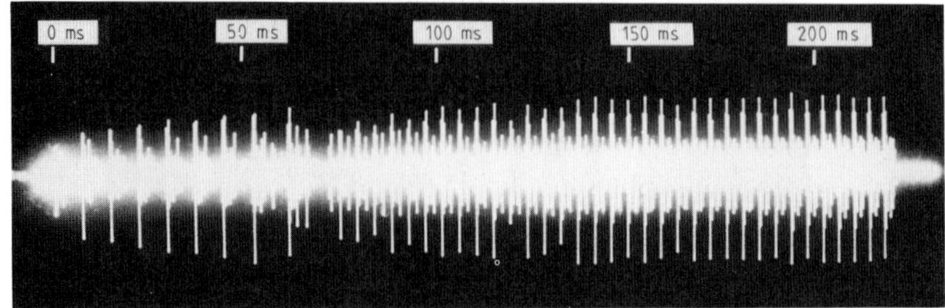

Figure 2.34 The trace of a note on a bagpipe chanter with the time duration in milliseconds. It can be seen that a steady state is not reached until about 150 ms after the initiation of the note.

2.20 MORE ABOUT THE ORIGIN OF TRANSIENTS

In section 2.6 we discussed the origin of the transient in relation to the placing of a struck tuning fork on a table. It was the inertia of the table that led to a characteristic delay as the table begins to vibrate. In section 2.11 we discussed the operation of the reed of a bagpipe chanter in maintaining the oscillation in the pipe. We did not, however, discuss the way in which the reed interacts with the pipe and, in particular, how the length of the pipe to the first open hole determines the ultimate pitch of the note produced.

The reed is virtually a tap which opens and shuts in order to let a succession of puffs of air, or compressions, into the pipe. Imagine the first compression about to start travelling down the pipe. It will travel until it reaches the first open hole and will then expand out through this hole. The air just above the hole will then expand and so on back to the top of the pipe. But the oscillation will be able to continue only if the expansion reaches the reed just at the moment when it is about to open to admit another compression. In section 2.4 we discussed the conditions necessary for the oscillation to continue and it is clear that unless the reed very rapidly adjusts itself (or is made to adjust by the player) to the frequency corresponding to the length of pipe in use, the oscillation will die away. But it takes a fraction of a second for this adjustment to occur and this is what gives rise to the very characteristic transient of the chanter. (It sounds almost like a very brief 'quack' as the note begins.) If a tape recording is made and then played back at about 1/16 of the normal speed the irregularities occurring during the transient can be clearly heard. Figure 2.34 shows the oscillograph trace of a note played on a chanter together with a time scale in milliseconds. The note starts at time zero but the oscillation is not regular until after about 150 ms.

2.21 CONCLUSION

In this chapter we have begun to study the ways in which the quality of musical sounds can be affected. In particular we have seen that most instruments do not produce pure musical tones and that the specific mixtures of harmonic components that occur can have a significant effect on the resultant perceived quality. We shall have much more to say in later chapters about the significance of the harmonic mixtures. But we have also seen that instruments do not produced continuous uniform tone. There is considerable variation in amplitude of the whole sound with time, there is a time variation of the relative proportions of the harmonics present and, perhaps most significant of all, the way in which the note begins, that is the starting transient, is one of the most important clues that help the ear–brain system to recognise a particular instrument quickly.

3

Science, Strings and Symphonies

3.1 INTRODUCTION

Stringed instruments like viols, violins, guitars, 'cellos, etc, rather obviously use long thin strings—i.e., one-dimensional things—as their primary vibrators and, as we saw in the last chapter, their sounds need amplifying before they can be used as instruments. The bodies of most instruments of this kind use a combination of flat plates—two-dimensional things—and enclosed volumes of air—three-dimensional things—to act as amplifiers and so we shall need to spend some time considering the vibrations of such objects in order to understand how the bodies of real instruments work. However, before doing that we shall look briefly at the historical development of the string family.

The true origin of the viols and violins has been the subject of great controversy for many years. It seems fairly certain that there existed in Ceylon (now Sri Lanka) as long ago as the third century BC a stringed instrument, played with a bow. It consisted essentially of a stick attached to a hollow cylinder across one end of which was stretched a skin on which rested a rudimentary bridge for the two strings. One theory is that the idea spread via India and Persia into Europe, but there are many other possibilities.

In medieval times there were many variations in Europe but quite how they were related to each other and to the ultimate survivors, the violins and viols, can only be a matter for speculation. Figure 3.1 shows some sketches based on medieval manuscripts and they pose some interesting questions.

1. Which hand is used for what? In (*a*) the player is bowing with his left hand but even though this leaves the fingering to the right *arm* it still seems to be a left hand!
2. Which way do you lean your head? In (*b*) the head leans to the right, and in (*c*) and (*d*) to the left.
3. Where do you bow? In (*b*) and (*c*) the bowing point is on the neck but in (*d*) it is near the tail piece.

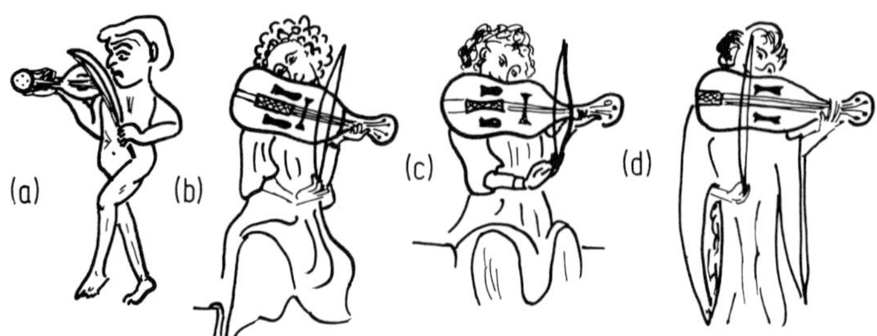

Figure 3.1 Sketches of string players based on medieval manuscripts: (*a*) the right arm appears to have a left hand!, (*b*) the head leans to the right and the bow is applied near the neck, (*c*) the head leans to the left and the bow is applied near the neck, and (*d*) the head leans to the left and the bow is applied near the tailpiece.

In the fourteenth and fifteenth centuries designs began to consolidate and there were two streams that began to emerge. The viols were at that time the more 'musical', and early violins, or fiddles, were regarded as rather crude instruments only suited to folk-type music for country dancing. Praetorius, the early 17th century musical historian, is not very helpful in tracing the various connections. In fact Boyden, in his *History of Violin Playing from the Earliest Times to 1761*, suggests that if only Praetorius had recorded his detailed knowledge of the state of the violin in the 16th century '... oceans of ink and countless hours spent in research might have been saved in the intervening centuries.'

By the end of the 17th century the viols, which were relatively quiet instruments more suited to chamber music, were being superseded by members of the violin family which, by that time, were capable of producing much more musical sounds which were sufficiently loud to permit the playing of symphonic music in concert halls. And Samuel Pepys quotes, in his famous diary, an anonymous ditty of the day

In former days we had the viol in
Ere the true instrument had come about.
But now we say, since this all ears do win,
The violin hath put the viol out!

The Golden Age of violin making was the first half of the 18th century but by the beginning of the 19th century the violin was still felt to be too weak to provide the power needed for the contemporary symphonies and concertos and practically all the existing violins including those of the great masters, Amati, Stradivari and Guarneri, were rebuilt. However, the question of why and how they were rebuilt will be more easily answered after we have looked at some of the fundamental acoustical principles.

3.2 PATTERNS OF VIBRATION OF PLATES

We saw in Chapter 2 that both strings and pipes have modes of vibration that are roughly harmonic. In other words devices in which one dimension is much larger than the other two have point nodes and approximately harmonically related partial frequencies. But as soon as we transfer to objects in which a second dimension is significant the modes are no longer even approximately harmonically related. The nodes also become lines instead of points. One of the best demonstrations of this is the experiment known as the Chladni plate, first performed by Chladni at the beginning of the 19th century. The following description is taken from my book *The Art and Science of Lecture Demonstration.*

> There are, of course many different ways of performing the Chladni experiment; the vibrations can be excited in a metal plate by means of an electromagnetic driver fed with a variable frequency sinusoidal current, or a wooden plate (e.g., the back plate of a double bass) can be excited by suspending it on foam pads over a large loudspeaker fed with variable frequency sinusoidal current. But my favourite uses Chladni's original idea of bowing, but with a metal plate. My own is about 20 cm square and is fixed to a 15 mm diameter pillar by a single screw through its midpoint. It is made of 1 mm thick brass and it is important that it should be cast plate, and not rolled, if symmetrical patterns are to be obtained.
>
> When I first started to perform this demonstration I was never quite certain what pattern, or note would be obtained and I would just touch the edge at various places while bowing and hope for the best. But gradually I found the trick of forcing a particular mode. For every different mode there are specific points along each edge where a nodal line intersects the edge and this is where a finger must be placed, and the bowing must occur midway between any two adjacent points. But unless you have several assistants to provide additional fingers it is impossible to cover every nodal intersection and there are often several patterns that have some intersections in common, and the selection of touching and bowing points needs to be made so that it is unique to one mode. For example the three modes shown in figure 3.2 all have an intersection at each of the four corners. To induce (*a*) on its own it is necessary to touch one corner (symmetry takes care of the other three); but then if you bow in the middle of a side there is ambiguity between (*a*) and (*b*); to remove this the bowing is done at the point marked p, thereby eliminating mode (*b*) which would require a node here.

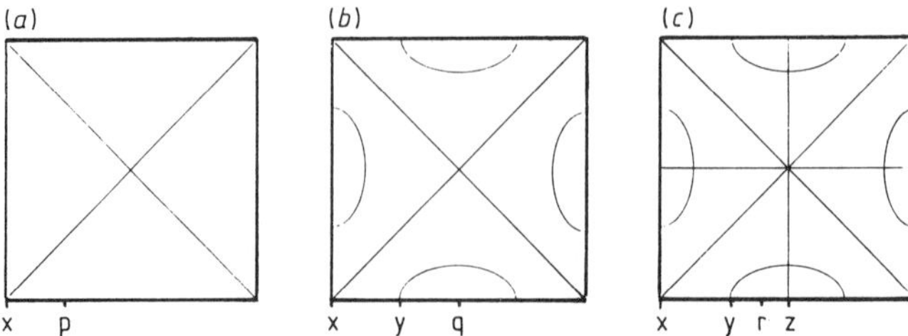

Figure 3.2 Diagram showing how three single modes on a bowed Chladni plate can be induced. Points x, y, and z are points at which a finger is placed to induce a node; the bow is applied at points p, q, and r.

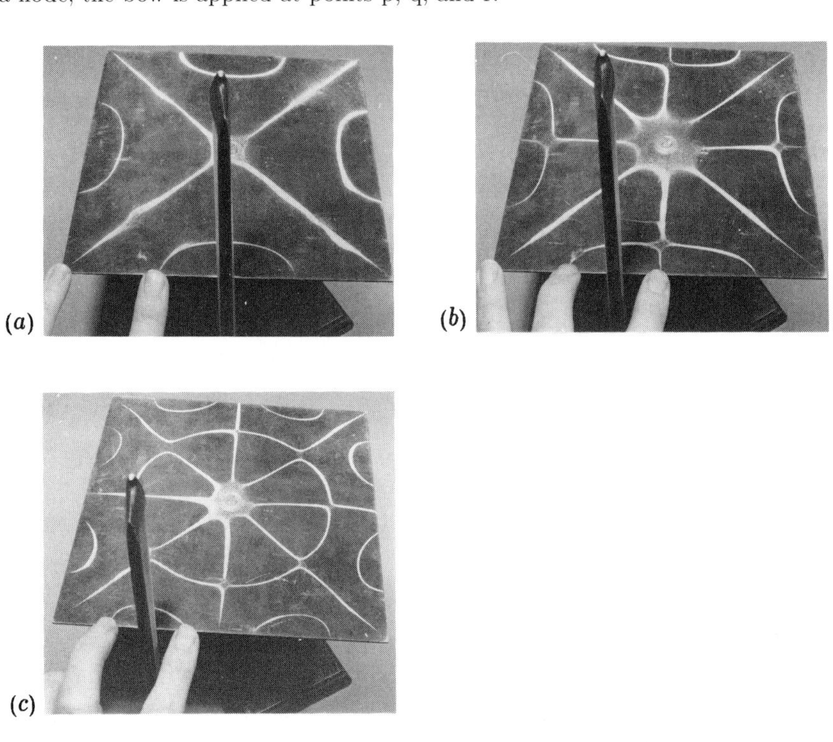

Figure 3.3 Photographs of the Chladni figures being produced: (*a*) and (*b*) correspond to figures 3.2(*b*) and (*c*).

To obtain (*b*) on its own it is necessary to touch at x and y to induce the two families of nodal lines and to bow at q in order to inhibit (*c*) which would have a node there. Finally to obtain (*c*) on its own three places must be touched, x, y, and z, and the bowing is at r.

(*a*)

(*b*)

Figure 3.4 (*a*) Repeat of the mode of figure 3.3(*c*) after lycopodium powder has been sprinkled all over the plate on top of the sand figure, and (*b*) the result of further bowing.

Figure 3.3 shows photographs of two of these modes and a third mode that belongs to a different series. Their frequencies are 1200 Hz, 2182 Hz, and 3828 Hz, respectively. There is obviously no easy harmonic relationship between them. Of course, if the plate is struck instead of being bowed, many modes are excited simultaneously. But the result is certainly not a harmonious sound; it is best described as a 'clang'.

Michael Faraday spent some considerable time studying the modes of plates under an enormous variety of conditions and using a great many alternatives to sand. In particular he tried lycopodium powder, a very finely divided and light material derived from the spores of certain mosses. He found that this powder behaved differently from sand; it collected at the *antinodes* of the pattern rather than along the nodal lines and showed that the violent vibration at the antinodes created air currents above the plate which carried the powder into the air; when the vibration ceased the powder fell back onto the plate. Figure 3.4(*a*) shows a plate on which the sand pattern of figure 3.3(*c*) has been recreated and then lycopodium powder has been scattered over the plate on top of the sand. Figure 3.4(*b*) shows the result when the plate has been excited again and the vibrations allowed to die away.

There are other ways of exploring the modes of vibration of plates. One very useful one is to excite the plate electronically and then to use laser holography to make the vibrational pattern visible. In order to excite a single mode the point of excitation as well as the frequency have to be adjusted. A very small cylindrical magnet is cemented at the point of excitation and then a coil fed with an alternating voltage at the appropriate frequency from an audio oscillator is clamped in position round the magnet.

A hologram of the vibrating plate is then made in the usual way. (Since holograms play such an interesting part in musical research a brief descrip-

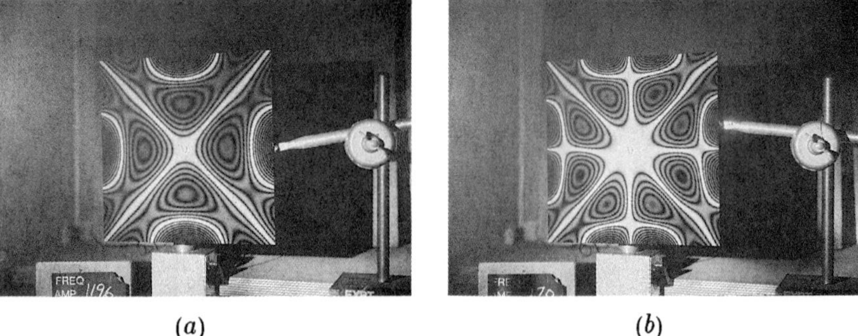

(*a*) (*b*)

Figure 3.5 Reconstructed holographic interference patterns for the modes of figures 3.3(*a*) and (*b*).

tion of the principles is given in Appendix A.) Any point on the plate *not* on a nodal line will execute simple harmonic motion about its rest position. It will therefore spend a much higher proportion of the cycle in one of the two extreme positions rather than in motion between the two extremes. During the photographic exposure lasting over a number of cycles there will, in effect, be two objects, one corresponding to each of the extreme positions and one coinciding along all the nodal lines. The amplitude of the motion is small and there will be interference fringes between the two reconstructed images which will reveal the pattern of plate vibrations as a series of contour lines. Figure 3.5 shows reconstructed holograms of two of the sand patterns of figure 3.3. The extra bright lines are the nodal lines along which no movement occurs and the relationship between the sand patterns and the reconstructed holograms is very clear.

It is quite difficult to envisage exactly how the plate is being bent during vibration, especially for the higher pitched and more complicated modes. Dr Bernard Richardson and his colleagues at the University of Wales College of Cardiff, who prepared the holograms, have also developed computer programmes to generate graphical simulations of the vibrations, and the resulting video sequences give a very clear picture of the movement of the plates. A still from each of two simulations is shown in figure 3.6 (they correspond to the sand patterns of figure 3.3(*c*) and (*d*)).

Amplification (or, more precisely, an increase in radiation efficiency) is not the only use to which plates are put in musical instruments. They can be used as instruments in their own right. For example the glockenspiel produces musical sounds from the transverse oscillation of metal plates when struck with a padded hammer. The celeste is simply a mechanised version of the glockenspiel in which the plates are struck by hammers controlled from a piano-like keyboard.

A rather more bizarre, but nevertheless interesting instrument is the musical saw. An ordinary tapered joiner's saw can be used, although 'con-

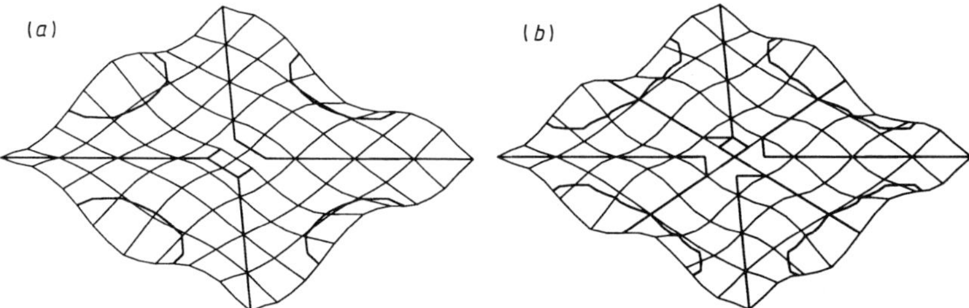

Figure 3.6 Stills from computer simulations for the modes of figures 3.3(a) and (b).

cert' saws are manufactured. If the saw is tapped it produces a nondescript 'tinny' sound. If it is bent into an arc of a circle as in figure 3.7(a) and tapped the resultant vibration is completely damped out. But if the saw is bent into an 'S' shape as in figure 3.7(b) it seems to 'come-to-life' and a light tap will produce a ringing vibration on a single note that will last for several seconds. Changing the bend near to the saw handle changes the pitch; the greater the degree of bending, the higher the pitch. A violin or 'cello bow can be used to feed-in energy (see section 2.16) and long sustained notes may then be produced. Provided that the 'S' shape is maintained the pitch of the note can be raised and lowered by increasing and decreasing the degree of bending to enable tunes to be played. The modal pattern may be revealed by holography as for the Chladni plate and the general form of the modes is shown in figure 3.8 The vibrations are clearly related to those of the Chladni plate and it is necessary to bow at the right place for each different note and this contributes to the difficulty of playing. (Obviously bowing on a nodal line will not work.) Even if the shape of the bend is held constant it is possibly to produce three of four different notes by exciting different modes. The angle of the taper turns out to be quite critical; without a taper it is difficult to produce a smooth glide from one note to another. It is not necessary to hold a finger on the saw as with the Chladni plate; it is sufficient to bow at different points along the edge.

3.3 PATTERNS OF VIBRATION OF AIR IN HOLLOW BODIES

One can imagine a continuous transition between a plate and a hollow body. If the edges of a circular plate are bent up it becomes a gong; further bending turns it into a bell; and finally the edges can be turned in on themselves to become a bottle. Figure 3.9(a) shows a large bronze

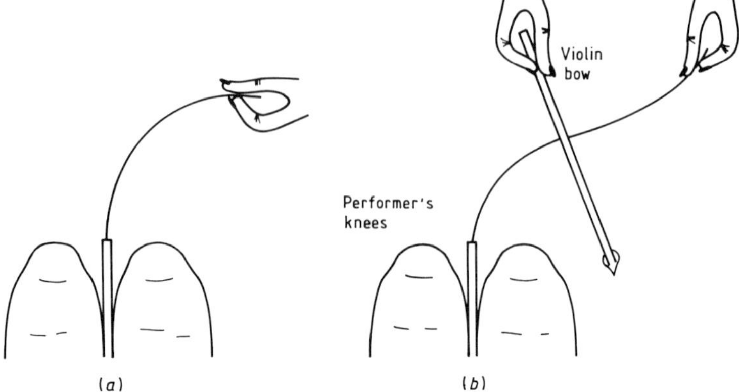

Figure 3.7 (*a*) If the saw is bent into a circle no vibration is possible. (*b*) Provided that the 'S' curve is maintained strong vibrations can be maintained and the pitch varied through changing the degree of curvature.

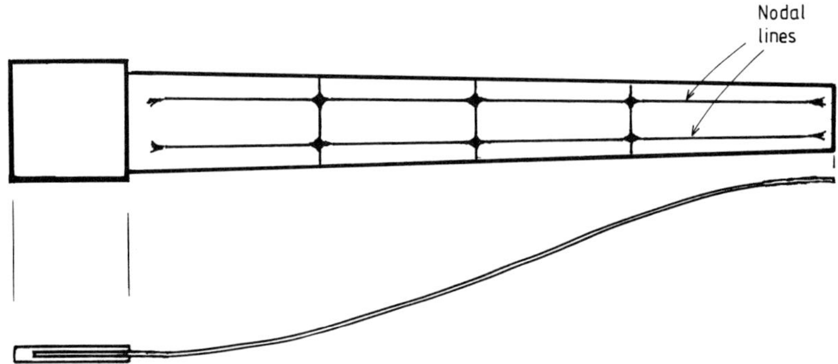

Figure 3.8 One mode of a musical saw sketched from a holographic interferogram.

bowl that John Tyndall used for many purposes. In the figure it is actually being bowed to excite one particular mode. If it is struck on the rim by a padded mallet it produces a set of ringing notes that persist for up to half a minute like a bell. Open and closed cylinders of variable length can be tuned to resonate and enhance one of the notes and the sound lasts long enough for the tube to be tuned in and out of resonance several times. One of the closed cylinders can be seen in the figure.

With objects of this sort of shape it is obviously difficult to find out anything about the nodal patterns using sand. But a small pith ball on a length of cotton can be used to plot the variation in amplitude of vibration.

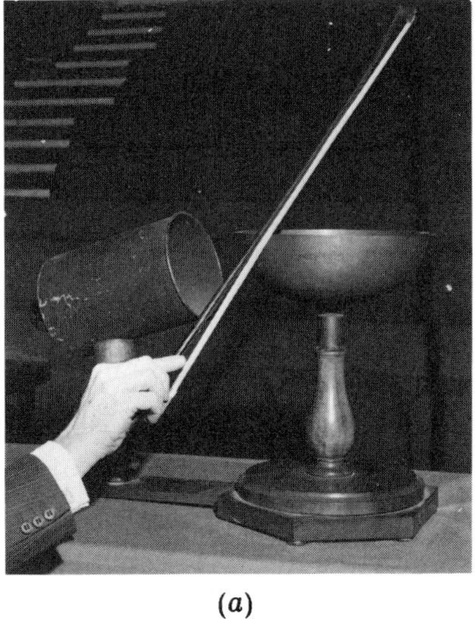

(*a*)

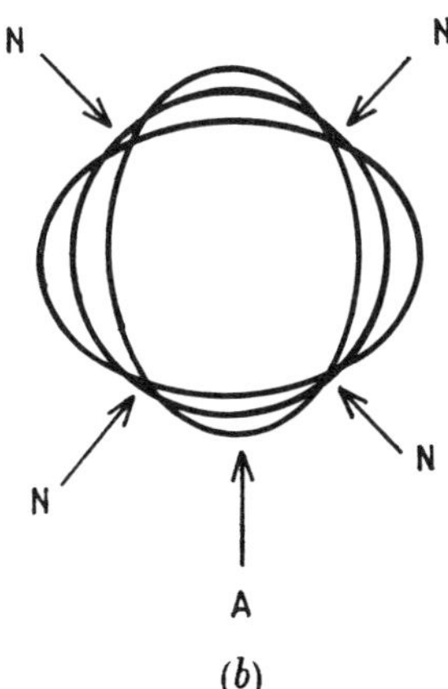

(*b*)

Figure 3.9 (*a*) Tyndall's bronze bowl being bowed at the Royal Institution. (*b*) Plan view of the pattern of vibration if the bowl is struck at A, which becomes an antinode. The nodes are marked N.

The edge of the bowl is struck and the pith ball allowed to just touch the edge. Where the vibration amplitude is large the ball is violently repelled, but at nodes there is scarcely any movement of the ball. It soon becomes clear that the bowl is vibrating in the pattern shown in the plan view in figure 3.9(*b*).

Just as with the Chladni plate, there are many different modes possible and, again, they are not harmonically related. The resonators Tyndall used with the bowl are designed to pick out the lowest note, the one corresponding to the mode illustrated in figure 3.9(*b*), but, if the bowl is struck without the resonator in position, a whole series of notes can be heard. The very characteristic tone of bowls and bells is quite different from that of devices with near harmonic overtones and it accounts for the special timbre produced by the bowls and gongs used in the gamelan orchestras of South East Asia.

The familiar water-filled wine glass stroked with a wet finger shows a pattern similar to that of the lowest pitched mode of a bowl, but now the finger stroking the rim plays a double role; first it acts like the fingers on the Chladni plate and eliminates all but the simplest mode and secondly it causes the whole pattern to rotate as the rim is stroked. It is important to notice that, although the excitation of the vibration by stick–slip motion (see section 2.16) is initially along the edge of the glass, this leads to distortion of the glass and to a similar pattern to that obtained by striking the glass. The note produced by striking and by stroking is, in fact, of the same pitch.

Adding water to the glass *lowers* the pitch of the note because it adds to the mass of the vibrating system. It is the glass and water that dominate the vibrations and the air in the glass plays little part.

However, if the bending of the plate is continued to form a bottle, then the air can play a more significant part. As with the wine glass, addition of water lowers the pitch of the note obtained by striking the bottle. But, if the air is excited by blowing across the neck of the bottle then the pitch of the note *rises* as water is added.

We now have what is, in effect, a Helmholtz resonator (as discussed in section 2.12) and it is important to remember that the resonant frequency depends both on the total volume and on the area of the aperture (in this case of the neck of the bottle).

The mouth behaves as a Helmholtz resonator and this can be demonstrated in several ways.

> If a tuning fork is held close to your open mouth it is possible, by adjusting the internal volume of the mouth to enable it to resonate to the tuning fork; this demonstration and the next one work best if a microphone and amplifier are used to pick out the sound coming from the mouth. An electric razor of the vibrating (rather than rotational) kind applied to the outside of the throat excites the mouth cavity with a sound that is rich in harmonics. With a little practice the size of the mouth cavity can be adjusted to pick out the harmonics in turn. The easiest way to find the right shape is to place the mouth in the appropriate shape for the enunciation of the vowel sounds.

Indeed we shall return to this idea in section 4.21 about the voice.

There are at least two instruments that are based on the Helmholtz

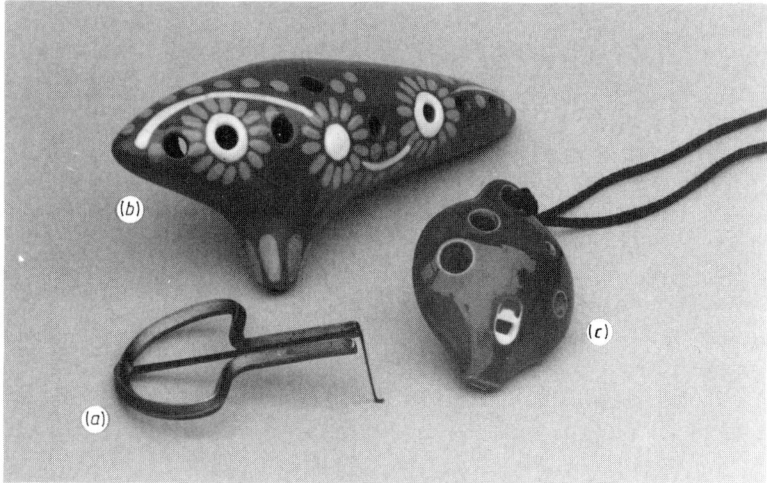

Figure 3.10 (*a*) A guimbard, (*b*) a ten-hole ocarina (two are thumb holes in the back), and (*c*) a four-hole ocarina.

resonator; one is the Jew's harp, or guimbard, and the other the ocarina. Figure 3.10(*a*) shows a guimbard in which the tapered tongue behaves very much like the musical saw. 'Twanging' the tongue produces a sound that is rich in modes, like hitting the unbent saw. The cavity of the mouth is then adjusted, as with the electric razor experiment, to pick out specific notes from the mixture.

Figures 3.10(*b*) and (*c*) show two different versions of the ocarina. The excitation is by means of edge tones with a mouthpiece resembling that of a recorder or tin whistle. But the interesting point is that, in a recorder or tin whistle, opening and closing the holes determines the vibrating length of the air column and therefore it is the exact pattern of open and closed holes that determines the pitch. In the ocarina, however, the pattern is immaterial; it is the total area of the apertures that matters and there may be any number of patterns of open holes that will give the same note. Usually, however, the holes are made of different sizes in order to achieve a given area with the smallest number of holes being covered or uncovered. The differences in size are particularly marked in the smaller ocarina (figure 3.10(*c*)). Although it has only four holes a full eight-note scale can be played.

3.4 THE BODIES OF STRINGED INSTRUMENTS

Now we can begin to consider the part played by the hollow bodies that are common to most stringed instruments. We will begin with the lute which can be traced back almost 4000 years. The essence of it is a hollow body shaped rather like half a pear to which is fitted a neck and a number of strings. It is played by plucking the strings and the pitch of a given note is determined by the vibrating length of a string which, in turn, is determined by the position of the fingers on the string. In Arab countries it was called the 'ud, or al'ud and that is possibly the root from which the name lute is derived, but research into the origins of names of musical instruments is notoriously hazardous and it would not be worth our while to spend time on those arguments in this book. There are dozens of shapes and sizes; the common features are the extremely light and hollow body and the 'frets' on the neck which help to locate the positions of the fingers in determining the pitch of notes. The frets were originally pieces of catgut tied round the neck, but in modern copies and in other related plucked instruments, they are usually strips of metal inset into the face of the neck. They not only help to locate the notes but also reduce the damping effect of the finger and allow the plucked note to be sustained for longer periods than would otherwise be possible.

The guitar is a similar instrument and may well have some common ancestry with that of the lute. But in the guitar the body is much larger and shallower and the modes of vibration of the large top plate are of much greater significance. The classical guitar has passed through many stages of evolution until it is now far louder and more tuneful and, in the hands of maestros like Segovia, has become capable of holding its own with a symphony orchestra in concertos. The construction of both the back and belly are by no means simple as they have elaborate systems of strutting which play a critical role in ensuring the desirable modal characteristics, and over which makers have their own very strong views.

My colleague Dr Bernard Richardson has made a detailed study of the vibrational behaviour of guitars and figure 3.11 shows a hologram of a finished guitar exhibiting a relatively high frequency mode. For the lower frequency modes the whole body is involved. In the lowest mode of all, the whole body expands and contracts and the front plate moves as a whole without any nodal lines except at the ribs. Figure 3.12 shows stills from two computer simulations (also due to Bernard Richardson) of the same kind as were discussed for the Chladni plate in section 3.2). Figure 3.12(*a*) represents the lowest mode and it is easy to see how air is pumped in and out of the body with the Helmholtz resonance of the body cavity playing a very significant part. In fact the frequency of this resonance depends chiefly on the volume of the body and the diameter of the hole. Since the whole of the front is moving in and out this mode is a very effective radiator of sound.

Figure 3.12(*b*) shows a mode in which one half of the front plate moves out while the other half moves in. This produces very little pumping action since the volume is much less changed than for the mode shown in figure 3.12(*a*). It is also somewhat less effective as a radiator because the two halves are in antiphase.

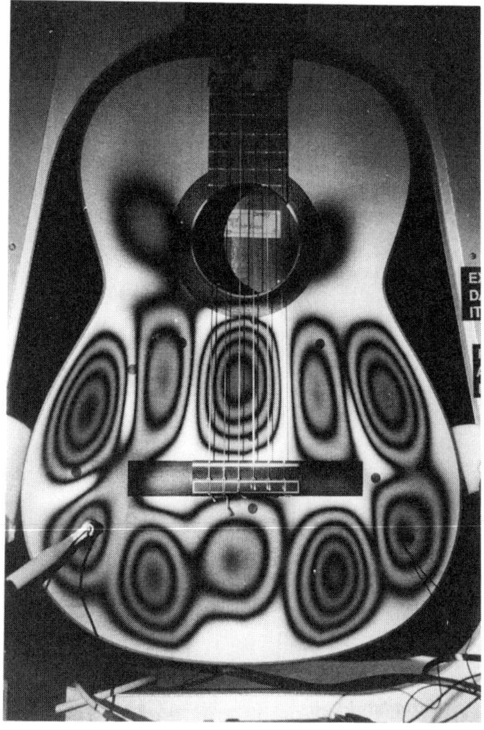

Figure 3.11 Holographic interferogram of the front plate of a complete guitar vibrating in a fairly high frequency mode.

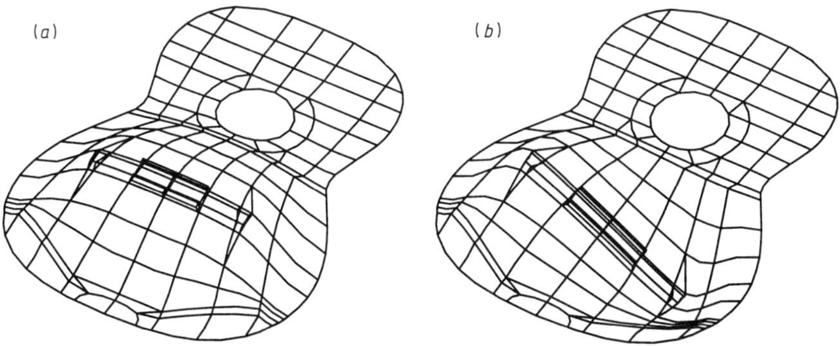

Figure 3.12 Stills from computer simulations of two low frequency modes of a guitar plate.

One of the questions that is a source of endless discussion and argument is that of the best modal frequencies for the front and back plates and the

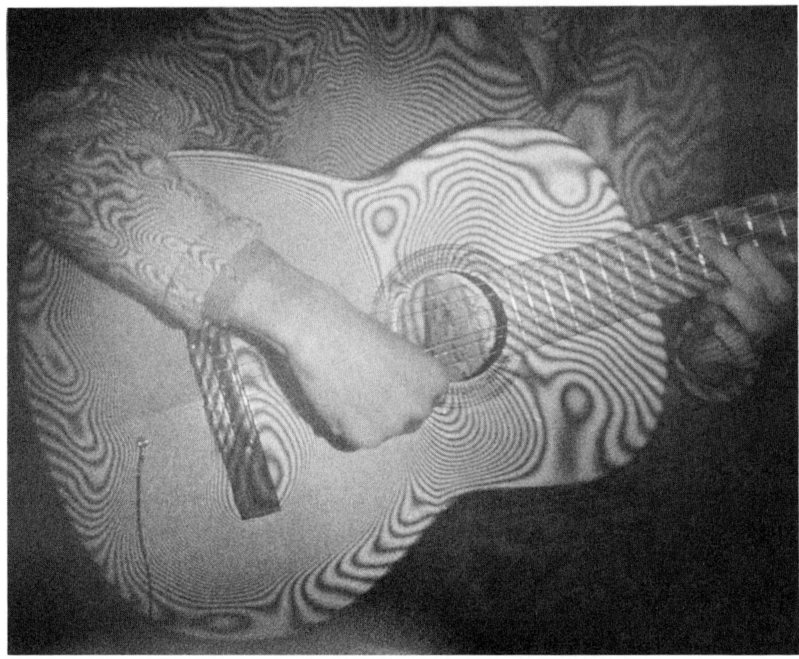

Figure 3.13 A real-time holographic interferogram showing vibration of the player as well as of the instrument. (Photo by courtesy of John Tyrer of Loughborough University of Technology.)

resonant frequency of the cavity. Although there are clearly substantial differences between the requirements of guitars and those of violins, many of the techniques used are so similar that it will be convenient to treat them together when we come to talk in detail about violins.

Before leaving the guitar there is one point that should be raised, particularly because it applies in some way to almost all instruments. Figure 3.13 is a photograph taken with a special holographic camera developed by John Tyrer at the University of Loughborough. It produces patterns of the same types as those of figure 3.11, but with the important difference that they can be produced in real time with a modified television camera. It is therefore unnecessary to excite the object electronically and to process and reconstruct the hologram. The camera can look at an instrument held and played by an instrumentalist as in figure 3.13. The extraordinary thing is that not only can the vibrations of the guitar be seen, but also vibrations on the player's hand and body. It is clear that the player's body cannot be ignored in considering the behaviour of an instrument under real playing conditions. It has also been suggested that these vibrations may relate to the remarkable ability of a few profoundly deaf, though very gifted artists to play various instruments at virtuoso standard.

3.5 BOWED INSTRUMENTS

There are many varieties of bowed instruments as was mentioned at the beginning of this chapter but for scientific purposes it will be convenient to concentrate on just three—the treble viol, the baroque violin and the modern violin.

The obvious features of the treble viol are that the instrument is fretted and has six strings like a guitar rather than the four strings of the violin.The body is much fatter than that of a violin. As mentioned at the beginning of the chapter its tone tended to be rather thin and weak compared with the violin and it was admirably suited to playing in a small room rather than in a concert hall.

The baroque violin is not fitted with frets and is superficially like the modern violin except that the neck is not tilted back, the shorter fingerboard is supported on a wedge and the bridge is lower. Most of the great violins of the Cremona school were not able to produce the volume of tone required by the symphonies of Brahms, Beethoven and others and were rebuilt between about 1830 and 1860. The result is that existing violins by Amati, Stradivari, and Guarneri, although having the original bodies, have been rebuilt to conform to the modern (i.e., post 1850s) configuration.

The drawings of figure 3.14 show the essential internal components of a violin. There are four strings tuned to be a musical fifth apart (G_3, 196 Hz; D_4, 293 Hz; A_4, 440 Hz; and E_5, 660 Hz). The strings are made of various materials. The first, or E string, is often made of silver steel, but catgut,

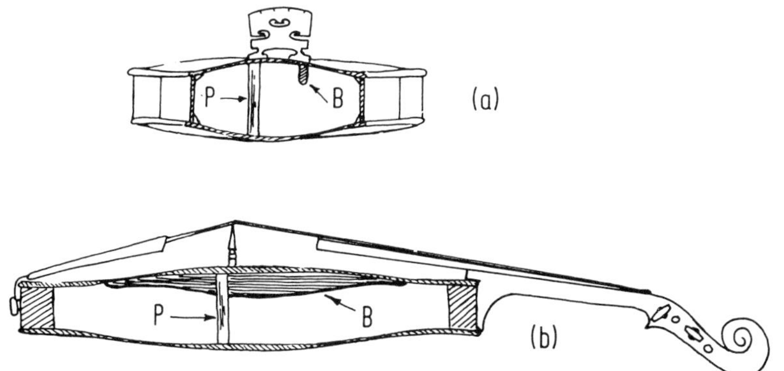

Figure 3.14 Transverse and longitudinal sections showing details of violin construction.

or even silk have been used. The second, or A, is often of gut alone, but sometimes is gut wound with very fine wire to add mass. The D and G strings are nearly always of wire wound on to a gut core. The manufacture and choice of strings is a matter of considerable debate and seems to be more a matter of personal preference on the part of the player than of real scientific differences. Tuning is by means of hard wooden pegs in the neck. In order to give a good grip to prevent slipping a mixture of chalk and powdered resin is sometimes applied. The strings pass over the bridge and are anchored to the tail piece. It is quite common to provide screw operated adjusters to aid fine tuning especially for the higher strings. The curve of the bridge is necessary to help the bow to play on any one string separately and the 'C' bouts in the side of the body are needed to clear the bow when playing on the lowest and highest strings. The finger board is tapered and curved and, ideally its upper surface should be close to, but obviously not touching, the strings at all points.

It sometimes comes as a slight surprise to people seeing an 'exploded' violin for the first time to find that the back and belly plates are not of uniform thickness. The variations in thickness can clearly be seen in figure 3.14. The 'F' holes play a very significant part in helping to determine the resonant frequency of the body; they behave like the neck aperture of a Helmholtz resonator. In this connection their total area is more important than their shape. But it is believed that the shape has another significant function in slightly weakening the coupling between the neck end of the belly and the tail end. Underneath the belly is glued a strip of wood known as the bass bar (B) and it seems that one of its many functions may be to compensate for the decoupling caused by the F-holes The bass bar also provides strengthening against the downward pressure of the bridge. The bass bar, as its name implies, lies more or less under the lowest string of the violin. At the other side, more or less under the foot of the bridge opposite to that over the bass bar, is the sound post (P).

There are endless arguments over the precise purpose of this. In one of the very early stringed instruments—the Welsh Crwth—the bridge was asymmetrical and one leg went down through a hole to rest on the back (see figure 3.15). It has been argued that this combines the function of the violin bridge with that of the sound post. Two things are certain; the first is that placing the sound post under one leg of the bridge helps to convert lateral vibrations of the strings into a rocking action of the bridge, which in turn helps the pumping action of the other bridge foot on the air in the body. The other is that the presence or absence of the post makes a significant difference to the tone of a violin. Its importance is reflected in the French word for the sound post—l'âme, the soul. Bernard Richardson has modified a violin so that the sound post can be removed and replaced from the outside in order to demonstrate the effect. Its lower end is fitted to a short piece of screwed brass rod and a matching screwed collar is fitted

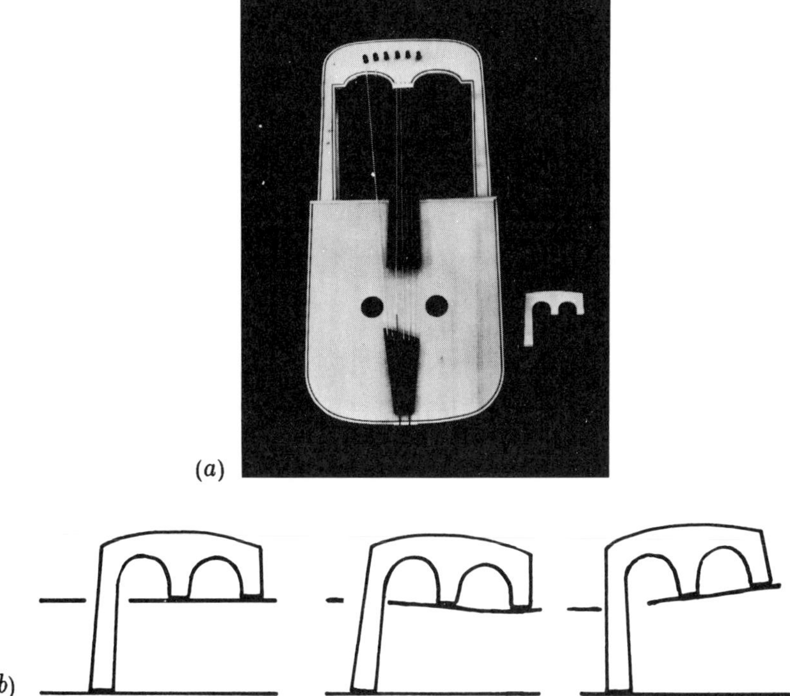

Figure 3.15 (*a*) Plan view of a Welsh Crwth with the bridge removed. (*b*) Shows how rotation of the bridge under the influence of string vibration causes a volume change in the body cavity.

to a hole in the back of the violin. The sound post is inserted through the collar and fixed in place by screwing the lower end into the collar using a special screwdriver that fits a socket in the brass end piece (figure 3.16). The instrument was demonstrated by Naomi Thomas with and without the sound post in position and the difference in the frequency analysis of the resulting tone could be seen on the real-time analyser, figure 3.17.

3.6 MAKING A VIOLIN

Each part of the violin is made of a different kind of wood and, of course, tradition plays a very large part in determining the choice. The two most significant choices from the point of view of the ultimate tone quality of the instrument, rather than the beauty of its appearance, are those for the back and belly. The belly is nearly always made of Norwegian spruce or something with similar close and straight grained characteristics. The back is usually of pear, cherry or maple and is made from two pieces of wood

Figure 3.16 Violin adapted by Bernard Richardson so that the sound post can be removed.

Figure 3.17 Naomi Thomas playing the adapted violin of figure 3.16 so that changes in the frequency responses can be recorded on the real-time analyser.

split from the same plank and glued together to achieve bilateral symmetry which has acoustic as well as aesthetic importance.

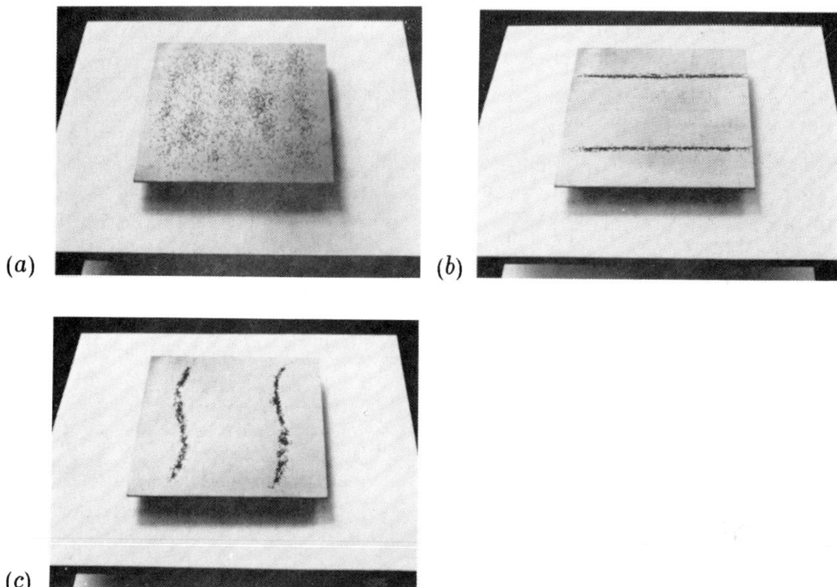

Figure 3.18 (*a*) A plate of spruce supported on small foam pads over a large loudspeaker and scattered with 'glitter'. The loudspeaker is driven with a sine wave of frequency (*b*) 91 Hz, and (*c*) 340 Hz.

For many years it was the custom for violin makers to adopt the precise dimensions of a famous violin—by Stradivari, Amati or another great maker—and to try to copy it precisely in every detail. Dimensions, variations in thickness, etc, were all slavishly copied and often, though not by any means always, the result was a rather second rate violin. This, we now realise is not a reflection of the skill with which the maker has reproduced the dimensions. The key point is that every piece of wood is different; even two pieces from the same tree will have differences in grain structure, elastic constants, damping constants and so on. It is therefore necessary—though obviously rather impractical—to redesign the instrument each time to match the specific characteristics of the wood used.

Some revealing demonstrations (again due to Bernard Richardson) can be done to show the significance of the grain of the wood in determining its behaviour from a musical point of view.

In the first demonstration two squares of spruce of the kind used for the front plate of a guitar are shown. In each the grain runs parallel to one side of the square. However, one of them is relatively difficult to bend across the grain and the other is very easy. The difference is that the stiffer piece has the grain perpendicular to the flat surface, and the more flexible one has the grain at 45° to the surface.

In the second demonstration a piece of spruce similar to the stiffer of the two used in the first demonstration is supported on small pieces of foam over a loudspeaker which is fed from a sine oscillator of variable frequency. Glitter of the type used in Christmas decorations is scattered on the wood (figure 3.18(a)) and the oscillator turned on. At quite a low frequency the pattern shown in figure 3.18(b) is obtained. To obtain the pattern of figure 3.18(c), the frequency has to be increased by a factor of almost 4 times. The significance is that the velocity of sound across the grain is only about a quarter of that along the grain. It has been suggested that this is one of the reasons why most stringed instruments that have a fairly flat belly are approximately 4 times longer than they are wide. Any vibration that is initiated in the middle of the plate will thus reach each side of the plate, travelling across the grain, at about the same time as it will reach each end travelling along the grain.

If elastic and other properties of the wood are so significant why then can we not use plastic which could have reproducible characteristics for as many violins as one cared to make? Plastic can be made with a grain by using carbon fibre reinforcement and, on the face of it, it would seem to be possible to design material with exactly the right characteristics. There seem to be two problems. The first is finding out what the ideal characteristics should be; and the second is that a violin would have to be completely redesigned from scratch to take account of the uniformity of the properties. So far, at least, success has not been achieved.

If you examine a violin or 'cello carefully you will see a double black line running all the way round the edge of both back and belly. This is called the 'purfling' and it is made by cutting a groove all the way round with a special tool and then inlaying a sandwich of two thin strips of ebony and one of a paler wood. Its original purpose may have been decoration; it seems to have a function in strengthening the edges and inhibiting cracks developing from the edges; but it certainly seems to have an acoustical effect as well, since it modifies the flexibility of the back or belly near the join with the ribs. It has even been suggested that, in a new violin the gluing-in of the purfling may set up some strains which may gradually be dispersed as the instrument is played. But there are so many strongly held views about the effects of aging, playing, etc, on the tone quality, at least some of which have little basis in fact, that I do not think it would be appropriate to comment further.

One of the key questions that faces all makers is the one mentioned above: it is necessary to alter the dimensions slightly to match the particular pieces of wood being used... but how? Violin making is a marvellous craft and one can only have the highest admiration for the skills exhibited by makers, most of whom have developed a whole range of 'tricks of the trade' to help answer the question. Holding the plates in various ways before they are assembled and tapping them at particular places, the so called 'tap–tone' test, is a widely used trick. The way in which the plate is held and the position of the tapping point determine the mode being excited in something like the same way as is done in the Chladni plate. It certainly helps to determine the frequency of certain modes of the plate. But of course the question still remains of what the frequencies should be? Then there is the question of how the frequencies of the modes will change when the instrument is assembled and the plates are glued to the ribs. And finally there is the question of the effect of the varnishing.

Everyone has heard the myth of the 'magic varnish' that is said to be the secret of the Cremona school and which will turn a mediocre violin into a masterpiece. Sadly it *is* largely a mythical notion. The real answer is that it would be better, acoustically, not to varnish at all! Any varnish will tend to dampen the vibrations. But if the instrument is not varnished it will change its qualities every time the humidity varies. The increase in humidity when a large audience enters a concert hall is enormous and consistent performance would be impossible. So the best varnish is the one that fulfils the sealing function with the least possible effect on the acoustic properties. Any heavy or sticky kind of varnish will obviously introduce extra damping and a hard, brittle varnish can change the stiffness of the plates quite a bit. So the varnish *is* of great importance but in a much less dramatic way than some of the myths would lead us to believe.

3.7 CAN SCIENCE HELP?

We said that the amplification by the body changes the quality of the sound and it must be clear that the real difference in behaviour between that of a cheap violin and of a Stradivari derives from the body. Can scientists perform measurements that will tell which of a number of violins is 'best'? It is certainly true that scientific tests can distinguish a bad violin from a good one. But it is still not possible to pick out the best from a group of good violins. In the first place we do not really know what is meant by 'best'. Many experiments have been done with players trying out a number of violins and it is quite difficult even to get agreement between players! One of the reasons is the way in which the brain works.

You may remember the experiments described in section 1.15 in which an audience finds it much easier to understand some distorted speech if it already knows what is being said. The same property of the brain is involved here. Each time a player listens to a violin the player's brain is being programmed. So each time an instrument is heard the sound is automatically compared by the brain with the sound heard on the previous occasion. If the listener is not the player, then slight changes in the player's reaction to different instruments may affect the listener's judgment. If the player is also the listener then it is impossible to disengage the 'feel' of the instrument, or the responsiveness and other factors from the objective assessment of the tone quality.

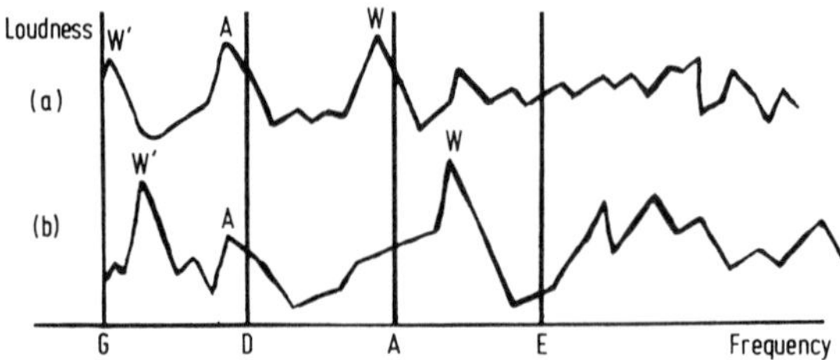

Figure 3.19 Saunders 'loudness' curves for (*a*) a relatively good violin and (*b*) a relatively bad violin. G, D, A and E represent the frequencies of the open strings; A represents the frequency of the air resonance of the body cavity; W represents the main wood resonance and W′ represents the frequency for which the second harmonic is at the wood resonance.

When a player picks up a violin for the first time, apart from the listening function, the brain is learning about the special peculiarities of the violin and how to get the best out of it. It may be several weeks before a real judgment can be made. So by the very nature of the exercise you cannot play first one and then another to get comparisons. So, even if there were scientific tests that could be used, trying to relate them with the opinions of players and listeners would be a significant problem.

What scientific tests or aids could be used? A great physicist—F A Saunders, who was also a string player—suggested a way of testing by getting a player to bow each note as loud as possible and then recording the loudness with a meter. A graph is then plotted of the measured loudness against the frequency This test will tell the difference between a very good and a very bad violin and typical graphs obtained by this kind of test are shown in figure 3.19. The bad violin is much less uniform as one goes from note to note. Clearly a player would have to compensate for these loudness changes and, quite apart from other factors it would obviously be a more difficult instrument to play.

So we might imagine that absolute uniformity would be the most desirable characteristic. It is not too difficult to achieve uniform loudness by using an electronic amplifier in place of the wooden body and the result is awful!

So we are presented with a philosophical problem that crops up in various guises throughout the scientific exploration of music. If Stradivari had been an electronic engineer would we now prefer uniform amplification? Without much hard evidence I would hazard a guess that we would not. I have a suspicion, built up from experience in many different areas, that the ear–brain system seems to prefer something that is not quite perfect. Perfection is too bland and we need a little non-uniformity to add a little bite.

Professor Max Matthews has developed a violin with electronic amplification. It has virtually no body—merely a metal frame which allows the player to hold it under the chin in the normal way. He has studied the variations in tone that can be obtained by altering the amplification at various frequencies and has found that an important component of the tone seems to be amplification at a particular relatively high frequency. He compares this with the so-called 'singing formant' which Sundberg suggested as the important feature that distinguishes singing from talking. This is discussed further in section 4.21 on the voice.

The Saunders loudness curve method of testing violins will certainly distinguish poor violins from good ones. But we are still no nearer finding a scientific method of distinguishing the outstanding violin from a group of good ones. One of the problems is that a good violin in the hands of a virtuoso performer becomes almost a part of the performer. The bow is an extension of the arm and an incredibly complex interaction is set up between the brain, the motor nerves and muscles in the player's arm, and the nerve signals that are sent back to the brain indicating the velocity, pressure, etc, of the bow on the strings (the proprioceptive response). I said earlier that it takes a while for a player to become used to a particular instrument and it is this kind of interaction that takes time to develop. But even a poor violin (in terms of its Saunders loudness curve) can sound quite good in the hands of a superb player. The player can even-out the large differences in response.

There is a legend that Kreisler was once billed to give a concert on his famous Stradivari violin. After the first piece which was applauded enthusiastically by the audience he broke the fiddle and then produced the real Strad. The story may be apocryphal, but, scientifically, the idea is perfectly possible. It is certainly true that a good player can make a poor violin sound very good. But it is also true that in doing so, so much of the brain is occupied with the problems of compensating for the irregularities of the violin that there is little left to deal with the purely musical problems.

Another test that has been used to determine the frequency response of the body is to drive the body using the same kind of electromagnetic actuator as was used to drive the Chladni plate for holographic interferometry (section 3.2). A good deal of experiment is needed to find the best position for the driver. The oscillator frequency is varied over the whole range and the audio output of the violin is measured by means of a microphone. Again the position of the microphone is quite critical as the direction of the radiation varies with frequency.

3.8 TESTING IN THE CONCERT HALL

None of the scientific methods so far described tackles the difficulty already mentioned in that the behaviour of the violin in a laboratory may be dif-

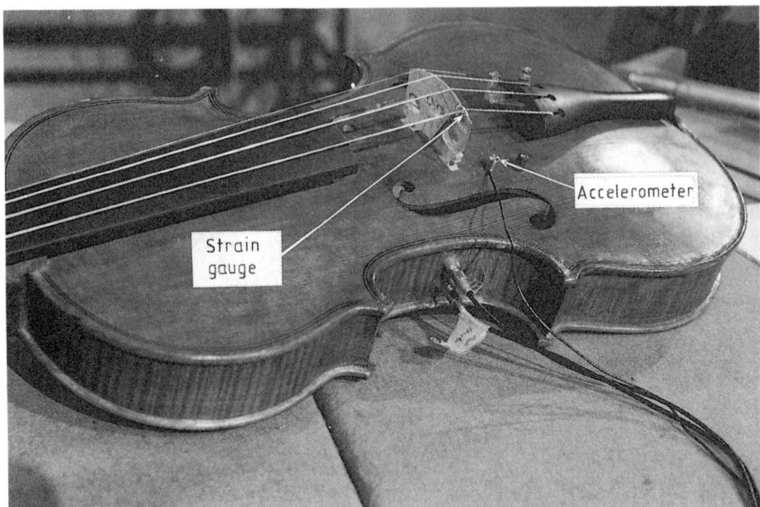

Figure 3.20 Violin with electrical sensors fitted by Bernard Richardson.

ferent from that in the concert hall, even when played by the same player. What is needed is some means by which the behaviour can be studied in the concert hall with the player facing an audience with all the adrenalin flowing.

In section 2.15 we discussed the basic principles of bowing and one of the many experiments that have been used to study the mechanism is one developed by Bernard Richardson. He uses a strain gauge attached to the bridge of the violin to monitor continuously the force exerted by the string on the bridge (figure 3.20). Figure 3.21 (upper trace) shows the oscillograph trace of the output from such a strain gauge. The way in which the string is first drawn to one side and then released can clearly be seen. The curve representing the working stroke has a series of spikes on it. These are almost certainly due to the fact that all the hairs of the bow do not release the string simultaneously and the spikes represent the behaviour of small groups of hairs or even of individual hairs. A moment's reflection on the factors that give rise to this behaviour may explain why good violin bows are so very expensive! The acoustic result of the spikes is the background bowing noise that can be heard, especially with an unskilled player.

If a small accelerometer is fixed to the body of the violin the response of the body to the forces applied by the bow can be monitored. (The accelerometer is, in effect, a magnet supported on springs inside a coil of wire. This is attached to the belly of the violin and generates currents that are proportional to the acceleration.) Its trace is shown in figure 3.21, middle trace.

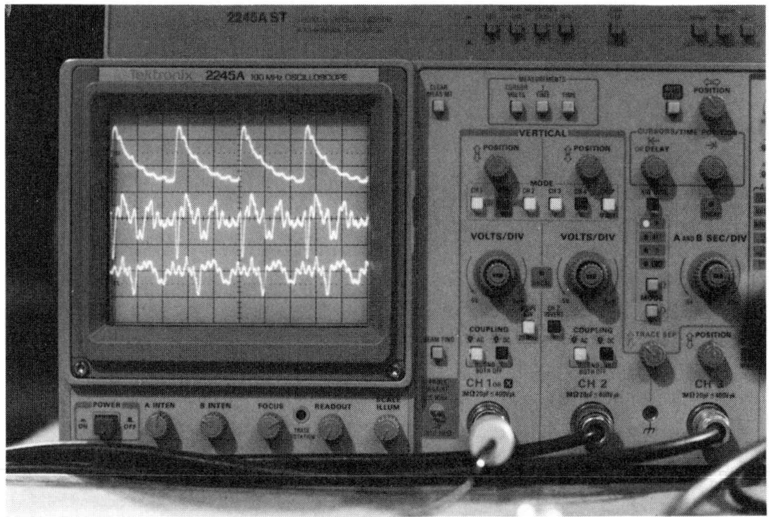

Figure 3.21 Simultaneous oscilloscope responses: upper trace, to the strain gauge on the bridge; middle trace, to the accelerometer on the front plate; lower trace, to the radiated sound as picked up by a microphone.

Finally the actual radiated sound can be monitored by a microphone system, and its trace is shown in figure 3.21, lower trace. The important point to be made here is that, though the upper trace varies little in general form from note to note, or, indeed, from violin to violin, there is enormous variation in the middle and lower traces.

But, even with such a system it is still difficult to find common factors between the best violins. A good deal of the problem lies in pinning down violinists to giving precise descriptions of the musical behaviour of their instruments and then relating them to physical effects.

For example violinists speak of the 'response' of a violin. They mean that some violins are quite difficult to play and others 'sing' the moment the bow touches the string. This is quite an elusive quality, but it certainly depends very much on the elastic and damping properties of the wood of the body. The varnish can have a big effect too. A thick sticky layer of varnish will certainly damp the vibrations considerably.

When a violin is being played the bow obviously starts up the string vibration and then continues to feed in the energy to keep it vibrating. When the player passes to the next note the vibrations of the body continue for a while until the next note is established. Thus the transient for a note in the middle of a sequence is different from that of a note played in isolation.

3.9 THE WOLF TONE

One of the problems that has puzzled violinists and 'cellists for generations is that of the wolf tone. On most violins and 'cellos there is one particular note at which the normal smooth tone of a good instrument seems to go out of control. The loudness of the note exhibits a kind of beat effect at a frequency of around 5 Hz and in the worst cases will sometimes jump an octave higher. The whole effect is reminiscent of those that sometimes occur when a choir boy's voice is beginning to 'break'. The literature is full of attempts to explain the phenomenon and it seems to have aroused a disproportionate amount of attention compared with other scientific problems associated with bowed strings. In violins the problem is usually circumvented by the maker attempting to place the wolf tone midway between two notes a semi tone apart so that it is unlikely to be used. But of course to place the wolf tone accurately means that the maker must have a good knowledge of the factors affecting its location.

The various explanations that have been put forward are reminiscent of the variety of theories of voice production which will be discussed in section 4.21; on the surface they appear to be utterly different and yet there are strong linking factors between them.

It seems to be agreed that what actually happens is that the fundamental frequency of a string being played by chance coincides with one of the natural resonances of the body that happens not to be very well damped. Most of us have seen the demonstration in which two identical pendulums are suspended from the same horizontal string. One pendulum only is set swinging and after a very short while the other comes into resonance with it but the process of energy transfer from one to the other goes on and eventually the second pendulum is swinging violently and the first has come to rest. The process begins all over again and the energy is exchanged from one to the other regularly until all is dissipated.

A similar phenomenon seems to occur with the wolf note; the string starts to vibrate and the body begins to take up some of the energy but then, just as with the pendulums, energy is transferred until the fundamental frequency vibration of the string dies away and all the energy is in the body. Then, of course the process is reversed. So the energy is swapped back and forth. When the fundamental has died away it is still possible for the second partial at the octave to continue vibrating and so the pitch appears to jump.

This is a very much simplified explanation and many complicated computer programmes have been written to try to simulate all the details of the phenomena that occur. Many attempts have been made to develop 'wolf tone suppressors'. The basic principle is to introduce a further system that will vibrate at the same frequency but which is heavily damped and will absorb the energy. Many of them make use of the portion of the string between the bridge and the tail piece. If this is heavily loaded the frequency can be adjusted and if the mass is made of some material that is not rigid (e.g., plasticene) sufficient damping can be introduced. It must be noted, however, that the suppressor has to be adjusted critically to suit the particular instrument.

3.10 THE CATGUT ACOUSTICAL SOCIETY

Before leaving the fascinating subject of the violin it would be wrong not to mention the society which has probably done more to organise and coordinate scientific studies of the violin and related instruments than any other in the second half of the twentieth century. In the last section I mentioned F A Saunders and it is from his pioneering work that the society has grown. An assistant of his—Mrs Carleen Hutchins—has been the indefatigable secretary since the society started with just twenty members in 1963. It now has well over a thousand members world wide. The members include musicians, instrument makers, scientists and many others, the common element being an abiding interest in all aspects of the string family. The society now publishes a semi-annual newsletter which not only contains members' papers but also lists publications by members in other journals.

One of the major tasks undertaken by Dr Hutchins in the early days of the society was to design and build a new family of instruments based on the violin. In the days when the viol was the dominant instrument, it existed in at least eight different sizes and much music was written for the 'Consort' of viols. The New Family of Violins is, in effect, a consort in the same sense as that of the viols.

The individual instruments will be described a little later but a few general comments are worth making at this stage. First, it seems to me that one of the great features of the family is that the work that has gone into its design and creation has increased our knowledge of violin making and testing enormously and would have been worth while for that alone. The original intention of being able to encourage the production of music for the whole consort has not been wholly successful. One problem is that each member of the family is significantly different in dimensions from those of its nearest existing instrument and professional players were often faced with a dilemma; if they took up the new instrument alone there was not much music available, whereas playing the new instrument might seriously affect performance on the normal instrument.

Another problem was that the instruments have a great similarity of tone in the areas where pitches overlap, whereas the conventional violin, viola, 'cello and double bass have different and very characteristic tone even in the overlapping regions. Thus the repertoire of pieces written for string quartets, etc, is not easily playable on the instruments of the new octet.

However, that having been said, many of the individual instruments of the family have excited considerable interest. For example the alto violin has great attractions for viola players and the baritone has a particularly rich tone much appreciated by 'cello players. We are of course venturing into a very subjective area where personal judgments and prejudices are possibly more important than scientific assessment.

In the early stages of the project a scaling theory was developed that would enable all eight proposed instruments to have their principal body and plate resonances to be at the same relative positions in relation to the string pitches. But it soon became apparent that if this were rigidly followed the instruments in the lower pitch ranges would be far too loud compared with the normal violin. It is well known that both the normal viola and 'cello are too small to give the richness of tone of which they are capable because, if made at a more appropriate scale, they become too large to be handled comfortably. But it was decided that the new family should be made to scale and that means should be found of overcoming the playing problem (as for example to play the equivalent of the viola vertically with a long pin like a 'cello).

To compensate for the loudness of the lower pitched instruments a new violin was developed (now called the treble violin) with larger top and back plates to increase the radiated sound, but with the arching, volume, etc, adjusted to bring the resonances back into the same relative positions as for a normal violin.

The complete family (see figure 3.22) will be described here because many points of scientific interest can be illustrated in passing.

The smallest instrument, the sopranino, is pitched an octave higher than a violin. Two problems arose immediately. The first was that if the scaling were strictly obeyed the instrument would be too small to be handled easily by an adult. It was accordingly made longer and the resulting low pitch of the air resonance was raised by making the 'f' holes very large. In our discussion of the Helmholtz resonator it was pointed out that increasing the area of the aperture raises the pitch. Even using the largest possible f holes the pitch of the air resonance was still not high enough and a series of holes was drilled in the ribs to raise the pitch still further.

The second problem was to find material of sufficient tensile strength for the E string. Eventually carbon fibre reinforced wire was found to be suitable but there remained an element of danger to the eyes if a string broke. The two solutions were either to wear safety goggles or to play the instrument resting on the forearm rather than under the chin in the way adopted by some folk fiddlers.

The descant is a musical fourth above the pitch of the violin, has a very sweet tone, and is one of the more successful members of the family. The treble is, of course at the same pitch as the normal violin but has a very uniform and powerful tone. The alto is at the same pitch as a viola and

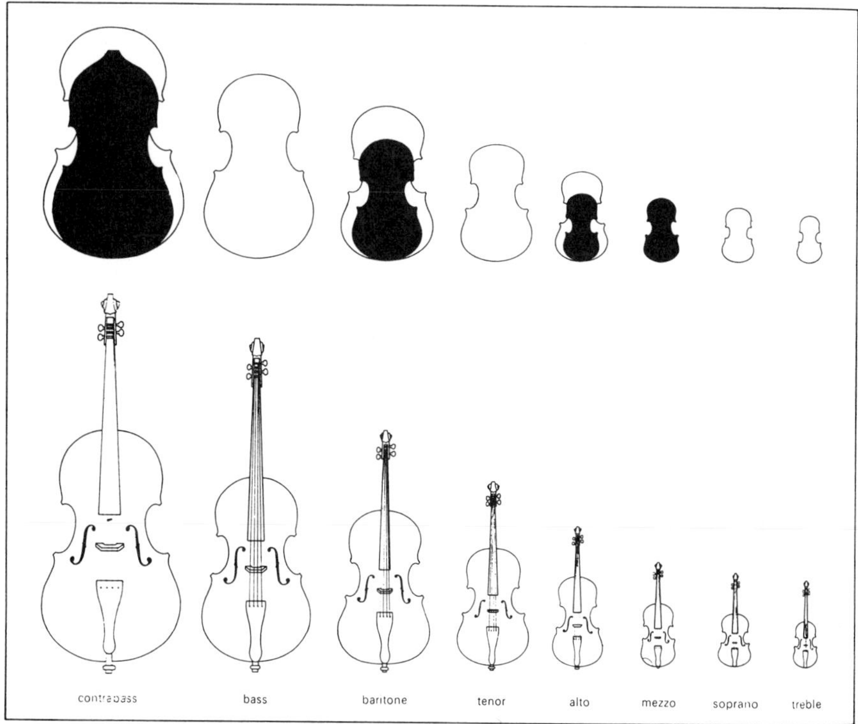

Figure 3.22 Sizes of the instruments of the Catgut Octet. The black shapes represent the outlines of the conventional orchestral strings on the same scale.

its rich and powerful tone has been greatly admired by viola players. It is considerably larger than a normal viola and this presents some playing problems as has already been mentioned.

The tenor is an octave below the violin in pitch and there is no equivalent among the conventional instruments. For some reason getting the resonances right for this instrument seemed to present greater problems than with any other members of the family. The baritone has the same pitch range as a 'cello and has a superbly rich tone. The remaining two are the small bass and the contra bass which are tuned in fourths like a normal double bass. The contrabass has the same pitch range as the double bass and the small bass is a fourth higher.

Several sets of the family have already been made including a number of additional single instruments. Competitions have been held for composers to write for the family and, though they have perhaps not fulfilled the original purpose of the project, they have, nevertheless, excited considerable interest and contributed greatly to our general knowledge of instruments of the bowed string family.

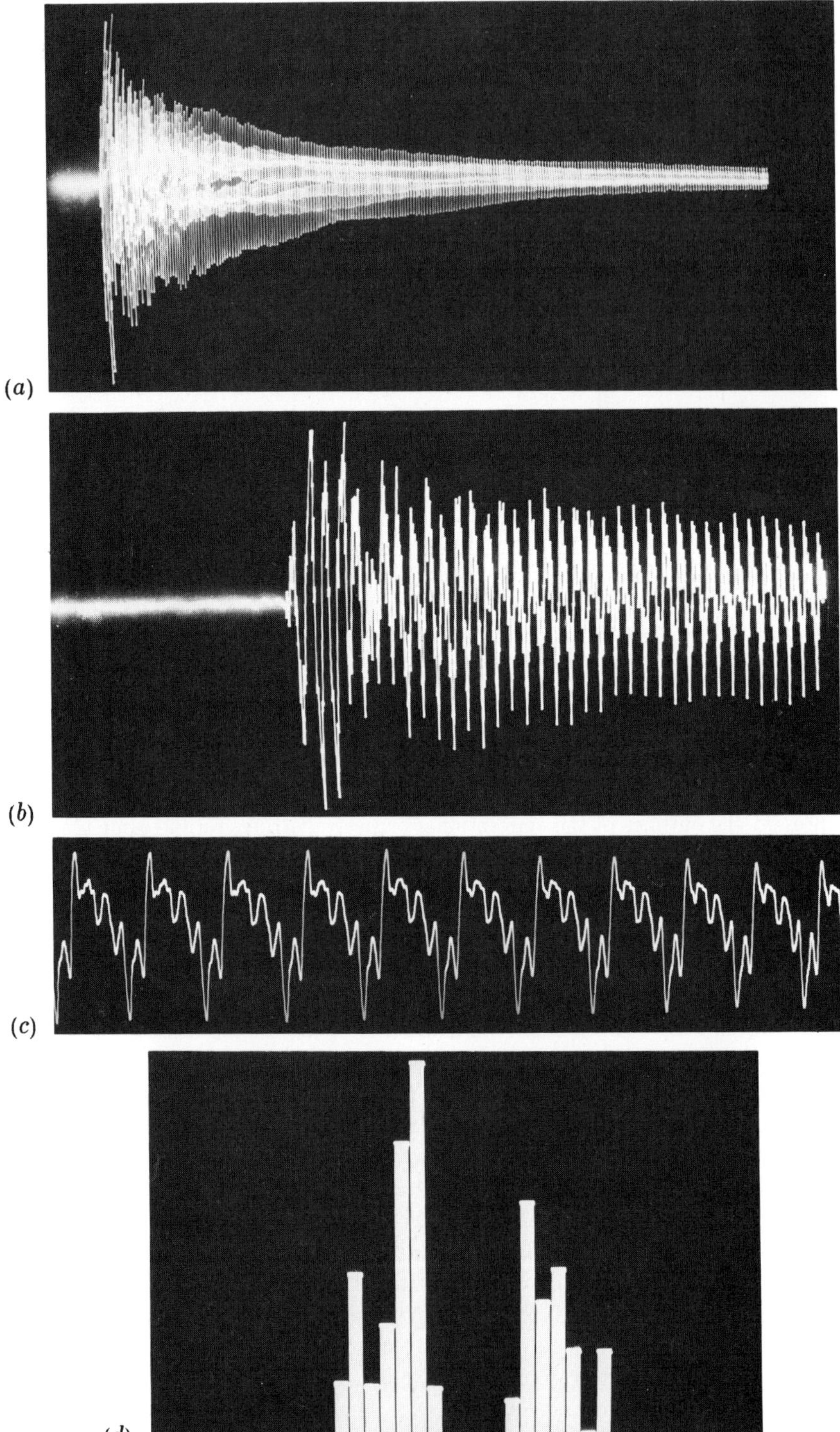

(a)

(b)

(c)

(d)

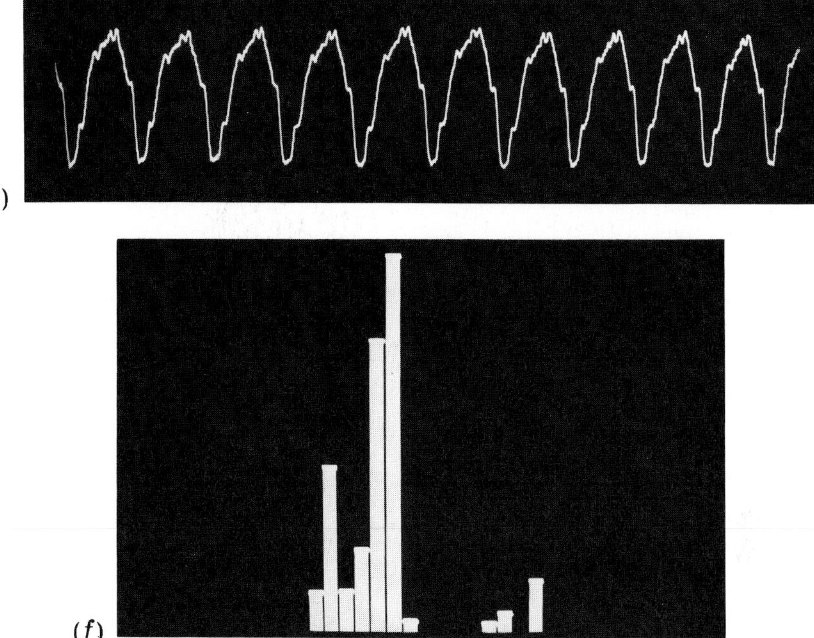

Figure 3.23 The top string (E_4, 330 Hz) of an acoustic guitar when plucked normally: (*a*) the trace of the whole note is displayed; (*b*) only the first 100 ms is displayed; (*c*) only a 30 ms length of note immediately following the attainment of maximum amplitude is displayed. (*d*) is the frequency analysis corresponding to (*c*), (*e*) corresponds to the string plucked with the broad ball of the thumb, displaying a 30 ms section as for (*c*), and (*f*) is the frequency analysis corresponding to (*e*).

3.11 HAND-PLUCKED STRINGS

We talked about the guitar in section 3.4 from the point of view of the behaviour of its body. Now we need to discuss the variation in quality that can be obtained by playing it in different ways. One of the most obvious ways of changing the quality is to use different objects with which to pluck the string. For example, in figure 3.23(*a*), we see the oscillograph trace of the whole duration of a note produced by normal plucking of the top string (E_4, 330 Hz) of an acoustic guitar while the other five strings are damped by wrapping felt round them. It is immediately obvious that the various components decay at different rates. In figure 3.23(*b*) we see the oscillograph trace of the same note as in figure 3.23(*a*), but lasting only 100 ms. In figure 3.23(*c*) the record is for the same note again, but a section only 30 ms long, just after it reaches the maximum amplitude, is recorded, and its frequency analysis is shown in figure 3.23(*d*). For

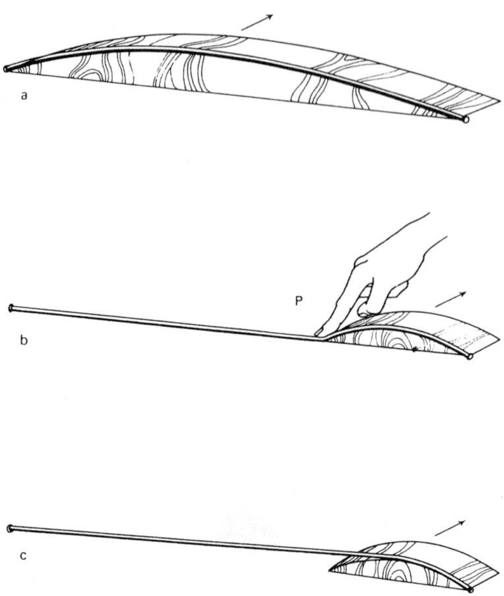

Figure 3.24 (*a*) A not-very-practical method of setting a string vibrating in its lowest mode by sliding the template in the direction of the arrow only. (*b*) Even less likely method of initiating the third mode alone. (*c*) Possible model of what happens when a string is plucked with the ball of the thumb.

the trace of figure 3.23(*e*) the broad ball of the thumb was used and the frequency analysis is shown in figure 3.23(*f*). There are obvious differences in the wave forms and analyses, and the corresponding sounds produced are significantly different. Of course, there are many other intermediate permutations that are possible. Why should such differences occur? In this book we are concerned with reasonable explanations rather than tightly reasoned arguments and I shall use the approach that I used in my first set of Christmas lectures (*Sounds of Music*). In section 2.15 we discussed the rubber cord as a model for the vibrations of strings. Look again at figure 2.28(*a*). Is it possible to think of a way in which the string could be set in motion instantly in this mode only? Suppose we cut a piece of wood to match the profile of the extreme position of the string and then used this to displace the string (see figure 3.24(*a*)). Then by moving the template sideways we could let the string slip off the edge. It is fairly clear that only the fundamental mode would be initiated. What we have done is to use the broadest possible plectrum to pluck the string and have produced only one mode.

Suppose we tried to excite only mode three of the rope by plucking? Again it might be possible to use the template method as in figure 3.24(*b*).

We should need to place a restraint at P and it is not entirely certain that we could excite only mode 3. Indeed if we did not place a restraint at P (figure 3.24(c)) it is likely that we would excite all of the modes 1, 2, or 3, since the shape of the cord at the end is very similar in all three. It is unlikely that we should initiate any modes higher than the third since these all require nodes at points in between P and the end of the string. But the template would force the string to move at all points between P and the end. In this way one can begin to see that in order to excite a great many of the higher frequency modes it would be necessary to use a very thin plectrum.

What about the point of plucking? We saw in section 2.15 (figure 2.28) that a particular harmonic can be encouraged by touching a vibrating string lightly at the position of one of the nodes of this mode. Conversely plucking the string at an antinode for this mode is likely to encourage this mode. Plucking in the middle of the string is therefore likely to induce only the odd harmonics, since these are the only ones with antinodes at the middle. Similarly, plucking at the middle will cause the whole string to vibrate and all the higher harmonics which would have nodes at various points between the middle and the ends would be inhibited.

The result of plucking somewhere near the normal position is thus obviously going to be that the string will vibrate simultaneously in a number of modes. In section 2.15 we talked about the ideas of Fourier summation and from what we have now said in this section, we can see that the various modes created by plucking with various widths of plectrum and at different points, will lead to different shapes for the composite wave. In figure 3.25 we can see the result of plucking a rubber cord at a point remote from the centre (i.e., the condition for a large number of harmonics) and the shape of the cord can be thought of as the Fourier summation of simultaneous vibration in a large number of modes. The series of flash photographs shows how the shape is maintained as the wave passes back and forth.

It is worth noting from figure 3.25 that if we consider the behaviour at any one point on the cord it will remain in one position (either that of maximum displacement upwards or of maximum displacement downwards) for large parts of the vibration cycle and then flip over as the wave passes along. At the midpoint for example the time spent in the two extreme positions would be equal. This is in marked contrast to the behaviour of a bowed string in which the 'flip' from one extreme to the other is different in each direction because of the stick–slip motion that is exciting the vibrations (see, for example, figure 3.21, upper trace).

3.12 KEYBOARD-OPERATED PLUCKED STRINGS

It will be convenient to consider instruments like the spinets, virginals and

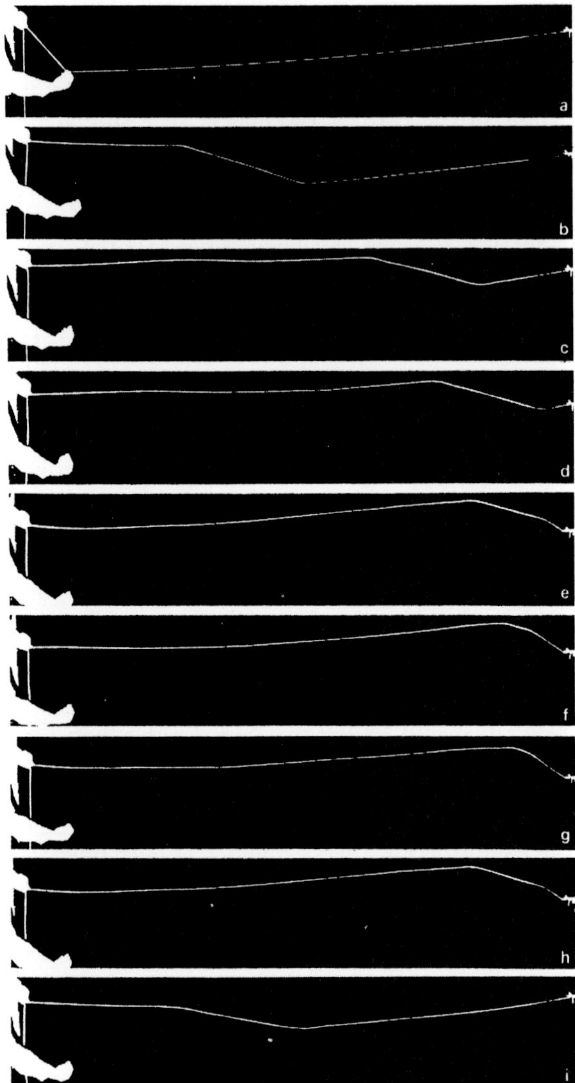

Figure 3.25 Instantaneous flash photographs of a wave travelling on a rubber rope that has been displaced as in (*a*) and then released. In (*b*)–(*f*) the wave is travelling from left to right and in (*g*)–(*i*) it is returning from right to left.

harpsichords at this point because the actual string behaviour is very like that of a guitar. The big difference, of course, is that these instruments belong to family number one (section 1.9) in which each string is virtually a separate instrument, whereas the guitar belongs to family number two.

The differences between these instruments are of scale and detail but

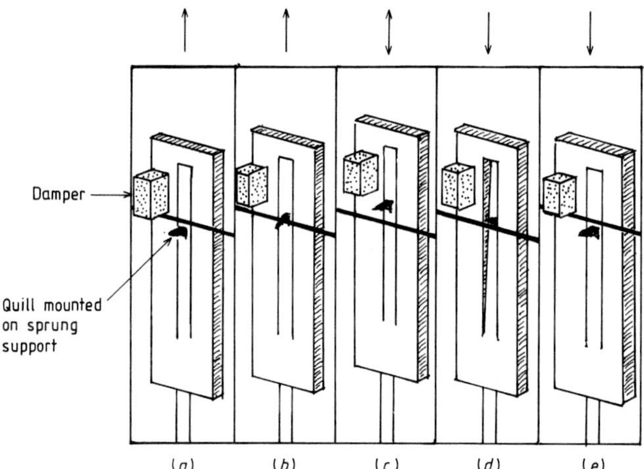

Figure 3.26 The mechanism of a harpsichord or spinet: when a key is depressed the 'jack' is raised and (*a*), (*b*) and (*c*) show its successive positions. In (*b*) the damper is raised from the string and in (*b*) and (*c*) the piece of quill or plastic, shown black, plucks the string. As the key is released, the jack falls back and at (*d*) the quill moves away from the string against a spring; in (*e*) the damper is replaced on the string.

all have a very similar mechanism. Figure 3.26 shows a diagram of the mechanism. When the key is depressed a piece of goose quill (nowadays often replaced by a specially formulated plastic) raises the string a little and then the stiffness of the quill is overcome and the string slips back. This provides the plucking mechanism, usually known as a 'jack'. The key also lifts a damper off the string to allow the vibrations to persist until the key is released. When the key is released the quill is kept out of the way of the string and falls back without creating further sound. The strings are mounted on a soundboard which provides the necessary increase in radiation to allow the sound to be heard.

There are many variations that can be added, but the basic principle is the same for all the keyboard-operated plucked strings. In the more elaborate harpsichords there are often two keyboards which operate two sets of strings that have the plucking devices at different points along the string to give different tone colour. The keyboards can be coupled together and often it is also possible to couple together keys that operate notes an octave apart, rather in the way that can be done on an organ.

Although a large harpsichord provides for differences in volume by the coupling systems just mentioned, the keyboard-plucked strings generally are not able to provide fine variations in loudness. Whether the key is struck gently or violently the loudness of a single note is unaffected. This is clearly because the string is always pushed the same distance from its null position and released. It is only with struck strings that we can begin to obtain fine variations in loudness.

3.13 KEYBOARD-OPERATED STRUCK STRINGS

The clavichord, though superficially similar to the instruments discussed in the last section uses a different principle. Its mode of operation can be demonstrated with an acoustic guitar. Place the guitar flat on a table and strike one string with the blunt edge of a knife and, having struck the string leave the knife in position to form a bridge on the string. Different notes will be produced depending on the position along its length at which the string is struck. Usually you will hear two notes which may or may not harmonise with each other. They represent the vibrations of the portions of the string on either side of the knife. If a piece of felt is wrapped round one end of the string the vibrations of one portion can be eliminated. The clavichord (figure 3.27) has a separate string (sometimes a pair of strings is used for each note to increase the volume) and a metal blade called a 'tangent' for each note. In order to save expense and space there exist instruments called 'fretted' clavichords in which the same string, or pair of strings is used for two or three notes. The neighbouring tangents strike the strings at different positions and each string usually provides three notes. The basic tuning is by tightening or slackening the end pins with a key. Tuning of the three notes on any one pair of strings is by bending the tangents.

With the clavichord, although it is essentially a very quiet instrument, considerable variation of loudness is possible by striking the keys gently or more vigorously. Also, since the tangent remains in contact with the string after the note has been produced a degree of 'vibrato' is possible by varying the pressure of the finger on the key.

3.14 THE PIANOFORTE

The pianoforte has a long and distinguished history both from the point of view of its evolution and in the development of playing techniques. It would be quite easy to fill several books with historical and technical details, but as far as this book is concerned I want to pick out one or two points concerned with the interplay between the design and the repertoire and about the various controversies that have occurred between scientists and musicians.

The action of a modern grand piano (figure 3.28(a)) must rank as one of the most brilliant pieces of mechanical craftsmanship ever to have been

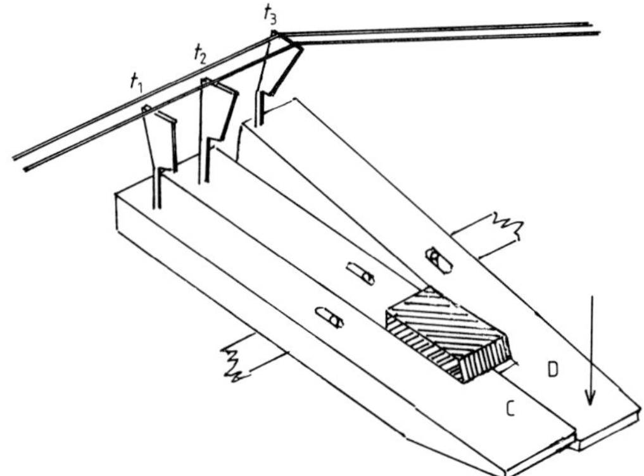

Figure 3.27 The mechanism of a 'fretted' clavichord: in this version one pair of strings provides three notes (C, C$^{\#}$, D). In the diagram the key D has been depressed. This action produces the note by striking the strings and also defines the length of string that can vibrate. Tuning is by bending the 'tangents', t_1, t_2, t_3, to the right or left. The left hand part of each string is damped with felt (not shown).

developed. Figure 3.28 also shows a demonstration model of the mechanism of one key. In (*b*) the mechanism is at rest and in (*c*) the key has been struck and remains depressed. The damper is still 'off' but the hammer has fallen back into an intermediate position ready for a second strike. To achieve its purpose every key must 'feel' the same to the player; every key must perform as nearly identically as is possible; every key must project the felt covered hammer to strike the string at a velocity that can be sensitively varied by changing the way in which it is struck; the possibility of the hammer bouncing and producing a double note must be eliminated and yet the whole system must be able to reset itself instantly so that multiple notes can be performed in rapid succession as in a trill, and damping action must be provided when the note is no longer required to continue sounding.

The whole of this complicated behaviour is achieved by means of a remarkable collection of wooden rods and levers, metal pivots, felt pads and leather strips that look like an inventor's nightmare and yet which perform their required functions beautifully. The soundboard of a modern piano is enormously strong and is supported by a massive steel frame. In addition to increasing the radiation from the strings to give the powerful tone that is one of its main features, it also imposes its own formant characteristics. The great strength of the frame and of the sound board permit the use of very stiff, thick steel strings, stretched to a very high tension and this is a further reason for the very loud tone compared with the earlier mechanised strings.

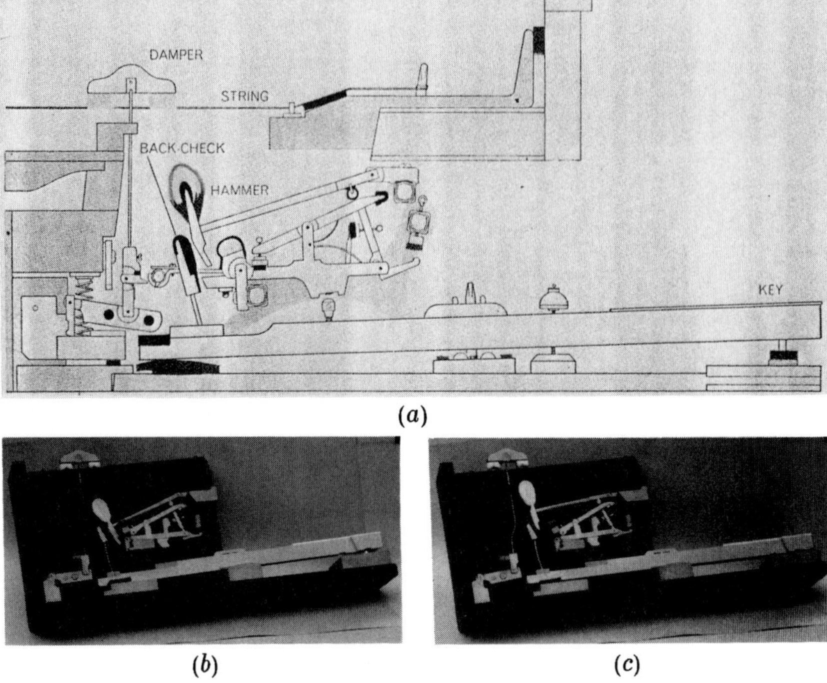

Figure 3.28 The action of a grand piano: (*a*) is a diagram of the main components with the mechanism at rest. (Reproduced by permission from *The Physics of the Piano* by E Donnel Blackham. © 1965 by Scientific American, Inc. All rights reserved.) (*b*) and (*c*) are photographs of the mechanism kindly loaned by Terry Pamplin of the London College of Furniture. In (*b*) the mechanism is at rest and in (*c*) it is shown after a note has been struck but with the key still depressed.

However, the thickness and stiffness of the strings causes their departure from the simple behaviour of the lighter and thinner strings of other stringed instruments. In particular the overtones are very non-harmonic. The first few are only slightly different from whole number multiples of the fundamental, but the fifteenth or sixteenth overtone may be as much as a semitone sharp from the true harmonic pitch. The mistuning increases rapidly with still higher overtones and in some reported electronic analyses reaches almost a fifth (three and a half tones) between the fortieth and fiftieth overtones.

One of the fascinating results of this is that, if piano tone is synthesised electronically with exactly harmonic overtones, the result does not sound like a piano and indeed is much less pleasant. If however the overtones are synthesised in the real non-harmonic ratios of an actual piano, the resultant tone acquires the quality usually called 'warmth' by musicians. This seems to be yet one more example of the fact that our brains seem to prefer some element of imperfection (refer back to section 3.7 on violin tone).

3.15 PIANO TOUCH

We must now turn to the controversial and fascinating question of 'piano touch'. Between the two world wars there raged a considerable controversy between physicists and musicians on the question. Physicists maintained (quite correctly) that since the hammer is not connected to the key at the time of impact with the string, i.e., that it is projected with a certain velocity by the mechanism, then the only parameter that the player can change is this velocity. The consequence of this is that dropping a brick on to a key, if the mass of the brick and the distance it is dropped are carefully chosen, should be capable of producing exactly the same quality as that produced by a pianist's finger. Naturally enough this excited considerable opposition from the musicians who claimed that there was all the difference in the world between the sound produced by dropping a brick and that produced by a pianist. Of course, as is so often the case in the fiercest controversies, the real point is that *like is not being compared with like.* The physicist's tests in the laboratory were concerned with single notes in isolation; the musicians were concerned with the playing of notes in a musical context. All kinds of variations then become possible.

One prominent feature of the behaviour of thick and stiff strings such as are used in the piano is that the partials decay at very different rates. By subtle variations in the force with which the keys are struck, the variation in the timing of the release of the dampers, by raising all the dampers using the sustaining pedal slightly before, slightly after or simultaneously with the striking of the key a great many variations are possible even with single notes. Varying forces on different keys within a chord, release of the dampers of the notes in a chord at slightly different times and many other subtle variations can easily add up to the remarkable package of effects usually lumped together under the designation 'touch'.

Much more recently Eric Clarke has performed some fascinating experiments on the effect on the listener of very small variations in the timing of the separate notes of a piece of music. He has devised computer systems that will calculate the minute departures from regularity in a performance and he very kindly performed these operations for me on the Nursery piece '*Ah! tu dirai-je Maman*'. Figure 3.29 shows the results of his experiments.

During the lectures, Yamaha–Kemble very kindly loaned us a 'Disklavier', which is a real piano (in our case an upright, but on a recent visit to Japan I was able to use a grand version). It has all the features of a normal piano and can be played normally. But, in addition, there are sensors which can record the precise movements of the keys, hammers, pedals, etc, on a 3.5″ computer microdisk. A series of electromagnetically controlled devices can then reproduce the performance, complete with all the mistakes, the variations in tone, tempo, etc. I suppose it could be described as an extremely sophisticated version of the old player piano. During the play back all sorts of variations are possible; variations can be made in the loudness, in the speed of the piece, the key can be changed and so on. The interesting point is that, even though the control is much more precise and subtle than even the best of the earlier mechanical pianos, repeated listening to the same piece shows that there are certain subtle characteristics that make it possible for an expert to distinguish between the original live performance and the recording. However, this should be interpreted as a tribute to the extraordinary powers of the human brain rather than as any criticism of the remarkable electronic mechanism.

3.16 CONCLUSION

In this chapter we have begun to consider in detail the problems involved in transforming the experiments on vibrating strings in an elementary physics laboratory into playable musical instruments. We have looked particularly at some of the factors that determine the tone quality of a finished instrument. We have also considered the extent to which scientific tests can be helpful and noted the significant differences that can occur between tests made under laboratory conditions and those made in the concert hall. Underlying all the differences in quality—which of course is a perceived quantity—lies the incredible human brain. Among other important abilities in the musical field is its power to distinguish between sounds that are made by machines and those in which a human being is involved.

Figure 3.29 *Opposite.* Eric Clarke's records of percentage departures from exact tempo in a 'one-finger' performance of '*Ah! tu dirai-je Maman*': (*a*) as indicated by the score, i.e., no deviations; (*b*) attempt by player to give a mechanical performance; and (*c*) expressive performance.

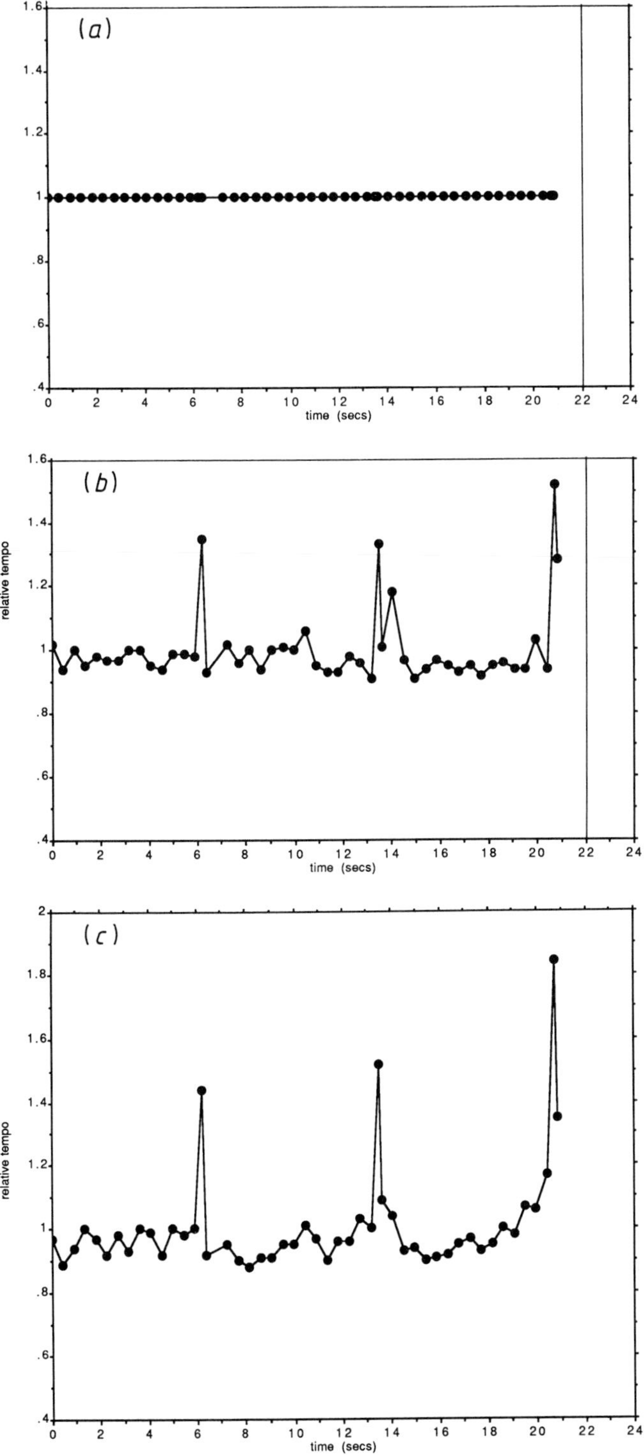
(a)
1.6
1.4
1.2
1
.8
.6
.4
0 2 4 6 8 10 12 14 16 18 20 22 24
time (secs)
(b)
relative tempo
1.6
1.4
1.2
1
.8
.6
.4
0 2 4 6 8 10 12 14 16 18 20 22 24
time (secs)
(c)
relative tempo
2
1.8
1.6
1.4
1.2
1
.8
.6
.4
0 2 4 6 8 10 12 14 16 18 20 22 24
time (secs)

4

Technology, Trumpets and Tunes

4.1 INTRODUCTION

In Chapter 2 we talked quite a lot about how oscillations could be set up in pipes, the modes of vibration of pipes and the use of reeds and edge tones to set up the vibrations and to feed-in energy to keep them going. In this chapter the main aim is to see what has to be done to turn a simple piece of tubing into a useful orchestral instrument. During the 20 or 30 years following the Second World War, sound was a neglected subject in many Physics departments, both in schools and universities. I am sure that one of the reasons was that the experiments usually performed in the practical laboratories (resonance in tubes, the sonometer, Kundt's tube, etc) seemed to be superficially related to musical instruments and yet, when the results from these experiments were applied to musical instruments, there seemed to be little correlation with the observed behaviour. For example, the odd sequence of harmonics could be demonstrated with the resonance tube experiment (cylindrical tube closed at one end) and yet a trumpet, which is cylindrical for most of its length and would seem to be closed by the lips at one end, plays a full series of harmonics.

I can certainly remember being very puzzled by this during my first year at university and it never seemed to occur to me that even adding a mouthpiece to a cylindrical tube, let alone the bell, was drastically altering the physical behaviour. There were many other similar problems and some led to the feeling that musicians and physicists had totally opposing views of what mattered in an instrument. This is the kind of problem that we shall be discussing and it will become clear that, nowadays, physicists, musicians and instrument makers can agree on the essential features that can turn simple tubes into splendid instruments.

It is also interesting that in several cases the development of the technology of instrument making, as for example valves in brass instruments, has had a profound effect on the repertoire of music produced by composers.

Although trumpets are the only instruments mentioned in the title of this chapter we shall in fact be discussing all kinds of wind instruments including not only both woodwind and brass, but also what could arguably be described as the most sublime of all, the pipe organ and the human voice.

For musical purposes wind instruments are divided into two main families, brass and woodwind. But, while there are obvious tonal relationships within the families, the names are inconsistent in many ways; for example, flutes are often of metal and yet are categorised as woodwind. It is more logical from a scientific point of view to divide them according to their method of excitation. For example edge-tone instruments (like the recorder or flute), wind-cap instruments (like the bagpipes or crumhorn) which use cane reeds that are not held in the lips, mouth reed instruments which use cane reeds that are gripped by the lips (like the oboe or clarinet), and lip–reed instruments, in which the lips form the actual reeds (like the trumpet or trombone). This is the division that we shall use in this chapter. But we shall begin by reminding ourselves of some of the features of simple tubes discussed earlier and develop them a little further.

In Chapter 2 we discussed the setting up of vibrations of air in tubes by sending a sequence of pulses into the tube and adjusting the pulse rate to coincide with a natural frequency of the tube. The simplest natural frequency of the tube is the one which corresponds to the time taken for a pulse to travel from one end of the tube to the other and back again once. The frequency corresponding to this is usually called the fundamental frequency but it can also be called the first harmonic. (The precise distinction between harmonics, partials and overtones is discussed in section 2.10.) In section 2.4 we noted that energy could be fed-in to keep a child on a swing moving by pushing once every second swing, once every third swing, etc. Now we should bring some of these ideas together to clarify the behaviour of vibrations in tubes before proceeding to discuss specific instruments.

4.2 WHAT HAPPENS AT THE END OF A TUBE?

We must first clarify what happens when a small pulse or compression sent down a tube reaches the end. A good demonstration model is a line of toy railway trucks connected to each other by springs running on a track. (To slow the demonstration down so that the behaviour can be seen more clearly I usually load the trucks with blocks of metal.) If the line of trucks is placed on a straight section of line and each end is left free this would correspond to a tube open at each end. A truck at one end of the line is moved-in towards the rest and released. The compression is passed from truck to truck until it reaches the last one. This truck moves off in the direction of travel of the pulse and, because there are no more trucks after it, the truck moves off until it is brought to a halt by the tension in the spring connecting it to the last but one truck. Thus the original compression is turned into an expansion and it is this expansion that travels back to the beginning of the line. Thus a compression sent along a plain cylindrical tube towards an open end will turn into an expansion on reflection. Conversely, an expansion sent along a plain cylindrical tube towards an open end will turn into a compression on reflection.

Another way to think about this is to consider the resultant pressure at the open end. This must be equal to the atmospheric pressure outside, and so the sum of the excess pressure in the forward travelling compression and the excess pressure in the backward travelling disturbance must be zero. Hence the backward travelling disturbance must be an expansion.

If the tube is closed at one end, then this can be modelled by placing buffers at one end of the row of trucks. Now when the compression reaches the last truck and comes up against the buffers the spring between the last but one truck and the last one will be compressed and this will send back a compression along the line. Thus a compression is reflected as a compression. It follows that an expansion will reflect as an expansion at a closed end.

4.3 VIBRATIONS IN TUBES OPEN AT BOTH ENDS

Let us first think about how vibrations at the fundamental frequency of a tube open at each end could be set up. The sequence of operations is shown in figure 4.1. The compression enters at the left hand end and travels to the right until it reaches the open end when it then travels back as an expansion. When it reaches the left hand end it will again become a compression and, if a second compression is fed in at precisely this moment the two will add and the amplitude will be increased. The oscillation can be maintained for as long as pulses are fed in at the frequency corresponding to the time taken for the compression to travel from left to right and back again. If we call this time T (the period of the oscillation) and if the velocity of the disturbances in the tube is v and the length of the tube l, then $T = l/v$ and the frequency $(1/T)$ is v/l.

How could such pulses be fed in? Any method that does not substantially disturb the 'open-endedness' of the tube could be used and the simplest way is to blow across the end as with the pan-pipes. The speed of blowing has to be adjusted until the edge-tone frequency is somewhere near the natural frequency of the pipe. Then the interaction between the edge tone and the returning pulses will pull the edge-tone frequency into tune.

If the speed of blowing is increased there comes a time when the note will jump up one octave. Now the edge tone frequency is roughly twice that of the fundamental frequency of the pipe, and there are two sets of pulses travelling up and down. The sequence is shown in figure 4.2. The pulse again enters from the left hand end, but a second pulse is now fed-in just

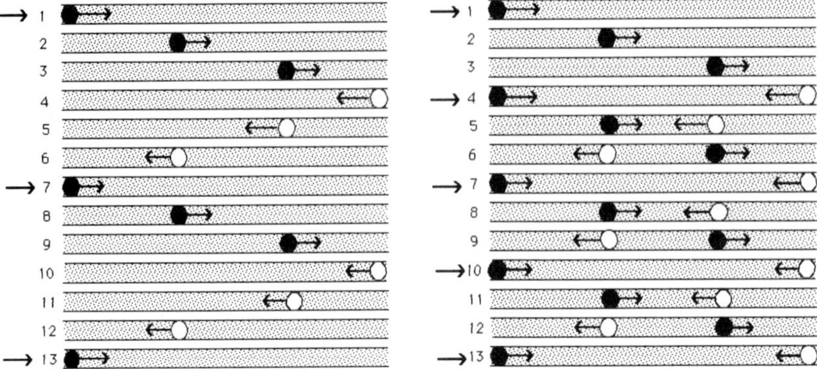

Figure 4.1 Sequence showing the build up of the fundamental in a tube open at each end. The black discs are compressions and the white discs are expansions. The time intervals between each successive picture are 1/6 of the period of the fundamental. To maintain the fundamental, pulses are to be fed-in at the times represented by pictures 1, 7, 13, etc.

Figure 4.2 Sequence showing the build up of the second harmonic in a tube open at each end. Pulses are now fed in at 1, 4, 7, 10, 13, etc.

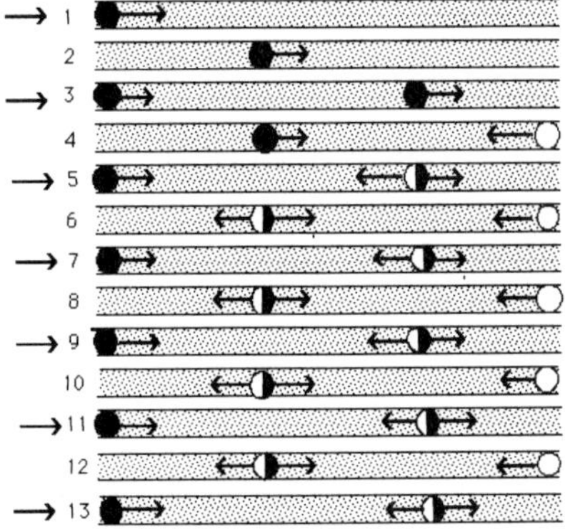

Figure 4.3. Sequence showing the build up of the third harmonic in a tube open at each end. Pulses are now fed in at 1, 3, 5, 7, 9, 11, 13, etc.

as the first one reaches the right hand end and its subsequent behaviour is exactly like that of the first pulse.

Similar behaviour occurs at three times the frequency, four times the frequency and so on. In fact a full series of harmonics can be produced and diagrams similar to those of figures 4.1 and 4.2 can be drawn for the other harmonics (figure 4.3 shows the diagram for the third harmonic).

Two more points are worth mentioning before we move on to tubes closed at one end. I mentioned earlier in this section that '...the interaction between the edge tone and the returning pulses will pull the edge-tone frequency into tune'. This is quite true; but it does take a little time. This provides one element of the characteristic starting transient (see section 2.20). If you make a tape recording of the note produced, say, by blowing across the end of a short length of drinking straw and play the tape back at about 1/16 of the speed, the note of course goes down four octaves. But the transient becomes 16 times longer and can be heard clearly as a variation in frequency as the note begins.

Finally, another way to initiate harmonics in a tube is demonstrated by the whirling corrugated plastic tubes that have been popular as scientific toys and can often be bought at 'hands-on' science centres such as, for example, Techniquest or Launch Pad. Hold the tube at one end and whirl it round. A sequence of harmonics can be heard and the faster it travels, the higher is the harmonic. But what is setting up the vibrations? It is quite complicated but we can get some idea of the mechanism by closing up the end held in the hand with a cork. Clearly air is drawn through the tube as it rotates. A simplistic view is that as the air rushes through the tube it hits the corrugations. If the rate corresponds with one of the natural harmonic frequencies of the tube, then the vibration will be maintained. Obviously the faster the tube is whirled, the faster will the air be drawn through and the rate at which the corrugations are hit will be raised.

Further demonstration of the fact that it is the flow of air through the pipe which causes the sound can be made with the apparatus being demonstrated in figure 4.4. The corrugated tube is sealed to the neck of an oil drum from which the bottom has been removed. When the drum is lowered into a dustbin full of water air is forced through the tube and its speed can be varied by varying the rate of movement of the drum. Sounds can be made either by lowering or by raising the drum.

4.4 VIBRATIONS IN TUBES CLOSED AT ONE END

We can use very similar arguments to those illustrated in figures 4.1–4.3 to find what happens if a tube is closed at one end. Figure 4.5 shows a cylindrical tube open at the left hand end and closed at the right. An initial pulse is sent down from left to right and the sequence of events is as shown.

Now, on reaching the right hand end, the pulse will stay as a compression on reflection and then, as shown in figure 4.5, it will become an expansion when it is reflected at the open end. So, in order to maintain the vibrations we would have to wait for the pulse to travel a second time to the closed end, be reflected, and then on reaching the open end again become a compression and be reinforced. Thus pulses have to be fed in at half the rate (assuming the same length of tube) as those for the open-ended tube. The note given is thus an octave lower.

Figure 4.4 Production of harmonics from a corrugated plastic tube connected at one end to an open-ended oil drum immersed in a tank of water.

Mathematically $T = 2l/v$ and the frequency $= 1/T = v/2l$.

When a child is learning to play the flute the teacher will often remove the main body of the flute and let the child blow across the hole of the top joint on its own. Very often too the open end will be closed so that the pitch of the note is lowered an octave. It then becomes nearer to the normal pitch of the complete flute and provides a useful way of enabling the pupil to concentrate on the lip and mouth shape and on breath control.

Now let us consider what happens if we feed-in pulses at twice this rate (refer to figure 4.5). Clearly the second compression will be due to enter the tube at the moment that the original compression has returned to the left hand open end and turned into an expansion on reflection. Obviously they will exactly balance out and so no second harmonic can be maintained. The third harmonic works satisfactorily (figure 4.6). In fact it turns out that none of the even numbered harmonics can be played.

4.5 PRIVILEGED FREQUENCIES

In the last two sections we have discussed what happens when pulses are fed in at a higher frequency than the fundamental for both closed and open

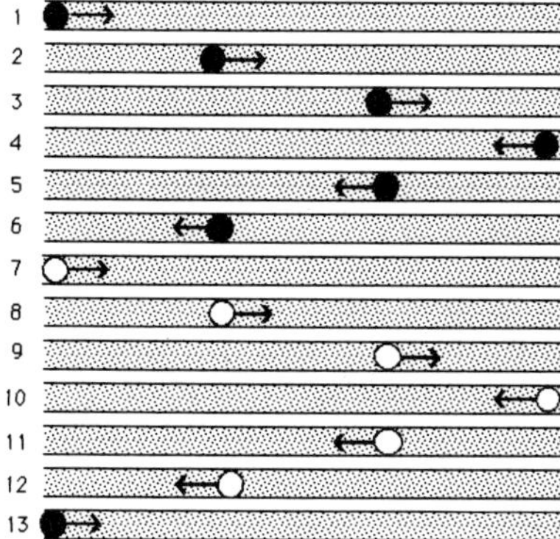

Figure 4.5 Sequence showing the build up of the fundamental in a tube closed at the right hand end and open at the left. The fundamental frequency is half that for the tube in figure 4.1. To maintain the fundamental the pulse is fed-in at point 1 and has to make a double transit of the tube before the next can be accepted at 13. If we try to induce the second harmonic by introducing a pulse after half a period, that is at point 7, it will clearly be out of phase and so no second harmonic is possible.

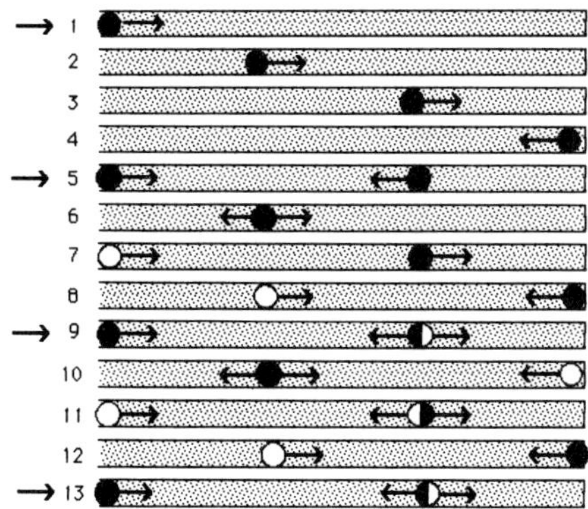

Figure 4.6 Sequence showing the build up of the third harmonic in a tube closed at the right hand end and open at the left. Pulses fed-in at intervals of one third of the period of the fundamental (i.e., at points 1, 5, 9, 13) will be in step.

tubes. Now we must consider what happens if pulses are fed-in at a *lower* frequency. For example, if for any given note (whether it is the fundamental or a harmonic, and whether the tube is open or closed) the frequency of the exciting pulses is halved, the excitation will continue. The parallel with the child on a swing would simply be that of pushing every alternate swing. It must be clear that the same could happen for every third swing, every fourth swing and so on. In other words, though the energy fed-in will be less and the excitation weaker, pulses at any sub-multiple frequency of any existing harmonic of the tube will keep energy fed-in. The great musical physicist, Art Benade, who died tragically young in 1987, used to call these privileged frequencies. The apparatus first introduced in Chapter 2, figure 2.12, to demonstrate how feeding pulses into a tube maintains oscillations can be used again to demonstrate the origin of privileged frequencies. Figure 4.7 shows the traces for two experiments; in the first, pulses are being fed-in at one quarter the natural frequency of the tube, and in the second at one sixth of the natural frequency. The tables of figure 4.8 and 4.9 give lists of some of the many possibilities for open tubes and for tubes closed at one end. These privileged frequencies explain why good brass players can often produce notes that are in between the expected harmonics and can also play notes that are far below the normal lowest note predicted by the simple theory. For example, a good trombone player can play a scale going *downwards* from the normal lowest note. The various departures from the behaviour expected on any simplistic theory were demonstrated in a particularly vivid way during the lectures by Dr John Bowsher. He is both an academic physicist and a very competent tenor trombone player. He first showed that it is possible to perform the characteristic trombone glissando, normally done by extending and retracting the slide, while keeping the slide absolutely stationary. He did it using his lips and, in effect, slipping through all the many privileged frequencies. The converse, keeping the note absolutely constant in pitch while extending and retracting the slide was, if anything even more astounding. In this case, as the slide moves, a different numbered privileged frequency is located and the pitch remains constant. I have seen the tape of this demonstration probably hundreds of times now and yet still find it almost unbelievable.

Finally he showed that even though the lowest note normally considered to be achievable on a tenor trombone is E_2 (82.5 Hz) he was able to play $E_2^\flat$ and continue downwards in the scale of $E^\flat$ and achieve $E_1^\flat$ as the final note!

4.6 VIBRATIONS IN CONICAL TUBES

We will now use the same reasoning that we used in sections 4.3 and 4.4 to consider what happens when pulses are fed-in to a conical tube. We will start with a tube that is not much smaller at one end than the other and will suppose that the pulses are fed-in from the small end. As the pulse travels along the tube there will be changes in the size of the disturbance as it expands along the tube but it is reasonable to suppose that, after

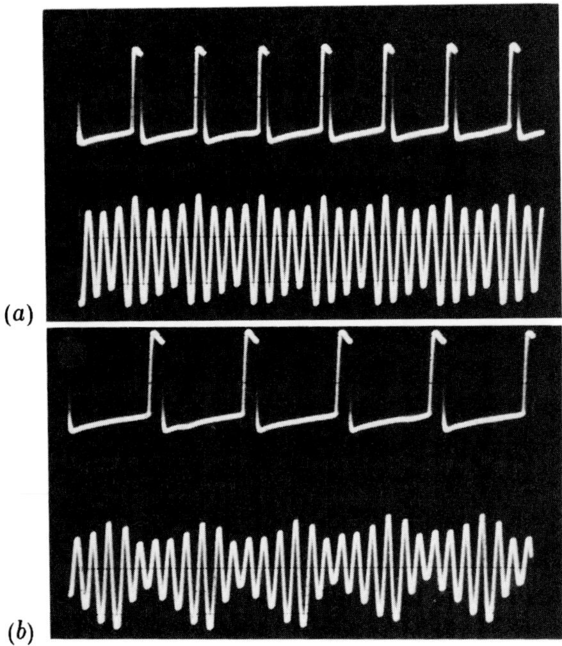

Figure 4.7 Traces produced using the apparatus of figure 2.12 to illustrate the maintenance of privileged frequencies. For (*a*) the pulses fed-in (upper trace) are at 1/4 of the fundamental frequency of the tube and the response (lower trace) shows how reinforcement occurs at every fourth transit of the tube. For (*b*) the frequency is 1/6 of the fundamental and reinforcement occurs every sixth transit.

240	480	720	960	1200	1440	1680	1920	2160
120	240	**360**	480	**600**	720	**840**	960	**1080**
80	**160**	240	**320**	**400**	480	**560**	**640**	720
60	120	**180**	240	**300**	360	**420**	480	**540**
48	**96**	**144**	**192**	240	**288**	**336**	**384**	**432**
40	**80**	120	160	**200**	240	**280**	320	360

Figure 4.8 Table of some privileged frequencies for a tube open at each end.

240	720	1200	1680	2160
120	**360**	**600**	**840**	**1080**
80	240	**400**	**560**	720
60	**180**	**300**	**420**	**540**
48	**144**	240	**336**	**432**
40	120	**200**	**280**	360

Figure 4.9 Table of some privileged frequencies for a tube open at one end only.

reflection at the open end it will regain its former dimensions by the time it is back at the left hand end. This qualitative reasoning can be confirmed by mathematical analysis. Thus, as far as the harmonics are concerned it will behave exactly as a plain cylindrical tube.

If we now imagine the small end of the cone to become smaller there is no obvious reason why the arguments should change. In fact, even when the small end becomes completely closed the behaviour remains the same. We thus reach the somewhat paradoxical conclusion that it makes no difference to the possible harmonics whether the small end of the conical pipe is closed or open! Again this conclusion can be supported by mathematical analysis. If you have difficulty in imagining what happens in this case it is important to remember that the dimensions and shape of the pulse change as it travels along the conical tube and it is these changes that result in what seems to be odd behaviour.

Thus we now have a reason why members of the clarinet family, which have cylindrical tubes with a reed which effectively acts as a closed end, have only odd harmonics possible when it is overblown. Members of the oboe family, on the other hand, have conical tubes and so, although the reed is virtually a closed end, have a full series of harmonics when overblown.

4.7 HARMONIC RECIPES IN THE WOODWINDS

It is important to notice that, so far, in talking about the harmonics of tubes, we have been talking about the behaviour of these instruments when overblown. The oboe family overblows at the octave, whereas the clarinet overblows at the twelfth (i.e., an octave plus a fifth). But we also need to consider the actual mixture of partial vibrations that are produced when the fundamental is being aimed at; this is one of the factors affecting the quality of the sound. In section 2.13 we mentioned Art Benades's term for the mixture of harmonics or partial vibrations making up a particular note—the recipe.

One important point that was mentioned in section 2.14 and needs to be stressed again here, is that the harmonic recipe for an instrument is not by any means the same throughout the pitch range of the instrument. The clarinet, for example, is usually said to have three distinct registers (though of course the transition between them is not a sudden one). The lowest register, corresponding to the fundamental mode of the pipe, is called by musicians the 'chalumeau' register. The notes produced at the first overblowing are called the middle or 'clarino' register and the notes produced by the second overblowing are the high register. A convincing demonstration of the differences in quality between the registers was described in section 2.14; it involved recording three notes an octave apart, one in each register. The recording of the high register note is then played back at a quarter speed, the note from the clarino register at half speed and the note from the chalumeau register played back at normal speed. The *pitches* of the notes then become the same and the differences in quality

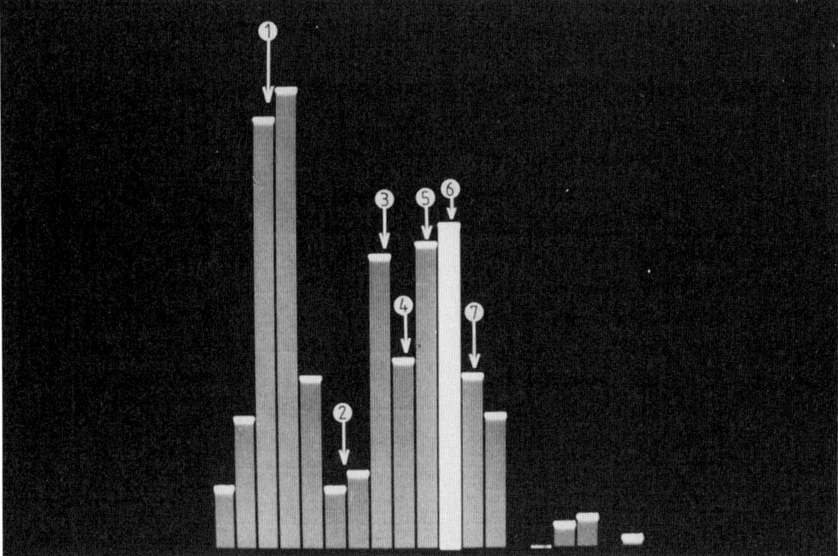

Figure 4.10 Analysis of the note F_4 (349 Hz) in the middle register for a clarinet. The waveform is the same as that for figure 2.27(*b*). The approximate positions of the first few harmonics are indicated.

become immediately obvious. Figure 2.26 showed oscillograph traces of such a set of three notes.

Figure 4.10 shows the analysis for the note F_4 (349 Hz) in the clarino, or middle register of the clarinet. Since the clarinet is in effect a cylindrical tube closed at one end, we should expect the even harmonics to be missing. Though not completely missing they are nevertheless much weaker than the odd harmonics. That they are not completely absent is one of the early points at which the behaviour of real instruments begins to diverge from what would be expected in terms of simple theory.

Among the reasons for the departure from simple theory are the fact that the reed is a non-linear device; that the tube is not plain but has a series of holes drilled in the side which, even when closed by a finger or a key, make discontinuities in the otherwise smooth cylindrical (clarinet) or conical (oboe and bassoon) bore; and that during the process of manufacture and of adjustment, minute changes in the bore diameter are produced, either by reaming at a particular point to increase the diameter, or by painting a thin coat of lacquer round the inside of the tube at particular points to reduce the diameter.

It can thus be seen that the real instrument is a far more complicated device than the simple tube found in the physics laboratory.

The recipe for the bassoon is particularly interesting. In figure 4.11 we see a bassoon being played and on the screen the spectrum is displayed on

Figure 4.11 A bassoon being played during the lectures with the frequency analysis, or harmonic recipe, being displayed on the real-time frequency analyser. Most of the energy can be seen to be in higher frequency components.

the real-time frequency analyser. The point to notice is that, although the note being played is in the lower register, most of the energy is concentrated in the higher harmonics (from about number 5 upwards). Thus although we may hear a note of apparent pitch, say A_2 (110 Hz), a great deal of the energy will be in notes such as A_4, $C_5^{\#}$ and E_5 (440, 550, and 660 Hz). How then do we hear the note A_2 (110 Hz), which is the note the bassoon is supposed to be playing? This is an example of the phenomenon that we called the 'missing fundamental' in section 1.16. The brain is, in fact, supplying the note of 110 Hz as the residue tone. In effect what happens is that because the notes at 440, 550, and 660 are successive harmonics of 110 Hz, the brain makes the assumption that there *must be* a fundamental at 110 Hz present.

Professor Schouten of the Institute for Perception in Eindhoven has produced a very elegant demonstration of the residue tone. The tune of the Westminster chimes ('Big Ben') at the hour consists of 16 notes. These were recorded but each of the 16 notes is double. The first note of each pair is a pure tone of the appropriate frequency, but the second note of each pair is a tonal complex consisting of a group of high harmonics of the required note which generates the required note as a 'missing fundamental'. Figure 4.12 shows oscillograph traces for the two types of note. The recording is then played back and low frequency noise is superposed. Immediately the pure-tone member of each pair disappears and only the

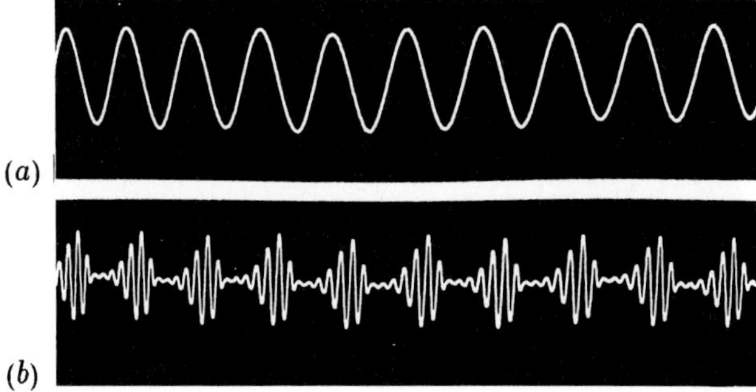

Figure 4.12 Waveforms for the two different notes used in Professor Schouten's demonstration: (*a*) is the pure tone and (*b*) is the tonal complex of high harmonics that seems to have the same pitch as (*a*) but with a very reedy quality.

complex is heard. The reason for this is the frequency-sensitive masking effect. The pure tone enters the aural network through a particular critical band (see section 1.10) and the low frequency noise is able to enter the same band and so masks it out. The tonal complex, which consists of high pitched sounds enters through a much higher critical band, is unaffected by the noise, and the generation of the fundamental occurs in the brain, rather than in earlier stages of the network. However, if the recording is played again and high frequency noise is superposed the effect is reversed. The pure tone is unaffected but the tonal complex is masked out. (A recording of this demonstration was issued as a supplement to the *Philips Technical Review*, Vol 24, 1962.)

4.8 EDGE-TONE INSTRUMENTS

The edge-tone mechanism was described in section 2.9. The alternation of the flow from one side of the edge to the other is relatively smooth and so the pressure change is more or less sinusoidal. As a result edge-tone instruments tend to give relatively simple, pure tones.

The earliest edge-tone instruments that still survive unchanged are the recorders. They are more or less cylindrical tubes, effectively open at both ends, and so will overblow at the octave. In order to play a normal scale in between the fundamental and the second harmonic we need to be able to shorten the tube in seven steps. This is provided by six holes (or in some cases seven) in the front which can be closed by the fingers and one hole at the back that can be covered by the thumb (see figure 4.13).

This basic fingering system is still at the core of most woodwinds, even those with the most sophisticated key work. In order to provide semitones there are various techniques. The uppermost open hole can be half covered,

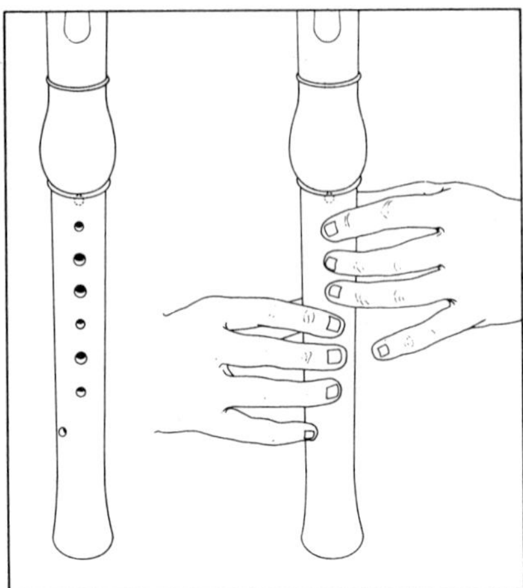

Figure 4.13 Finger hole positions on a descant recorder.

or this hole may be left open and the one or two holes next below it can be closed. The second octave and even higher notes can be produced reasonably in tune by overblowing but complicated systems of 'cross fingering' are sometimes introduced. In effect the pattern of open holes relates to the system of nodes and antinodes in the pipe and helps to maintain the higher modes of vibration.

The recorder family consists of five members: bass, lowest note F_3 (174.6 Hz); tenor C_4 (261.6 Hz); treble F_4 (349.2 Hz); descant C_5 (523.3 Hz); and sopranino F_5 (698.5 Hz). They are very popular instruments with children because they are relatively easy to play. Of course there is ample scope for virtuoso performance at a later stage, but at least an acceptable quality and performance reasonably in tune can be obtained almost immediately. There is really only one variable parameter and that is the wind pressure. Control of the attack can be performed with the tongue and rapid staccato passages can be produced by double and triple 'tonguing'.

The flute is often said to resemble the recorder but in fact there are many other differences in addition to the basic one of being blown in the transverse position. The main body is a tube that is cylindrical over most of the length with a bore of about 3/4 inch but at the mouth hole end it tapers down. The overall length of the standard orchestral flute is about 26 1/2 inch and the mouth hole is about 2 1/2 inch from one end. This end is closed by a cork whose position can be varied by an adjusting screw for tuning purposes. The mouth hole itself is made in a thickened part of the tube (or sometimes in metal flutes in a flange that is fixed to the tube). The design of the mouth hole is a matter of much controversy and still seems to be a matter of experiment with trial and error contributing more than physical theory.

The pitch range of a normal orchestral flute is about three octaves with middle C (C_4) as its lowest note. The piccolo is a flute with its lowest note an octave higher.

If the length of a flute is carefully measured and its fundamental pitch predicted it will be found to be at least a semitone higher than the note actually produced. The difference is accounted for by the additional volume of the 'bumps' in the tube under the closed pads on the tone holes and of the space between the mouth hole and the tuning cork.

Further deviations from simple pipe behaviour arise because the lips of the player are slightly hooded over the mouth hole and this is one of the important parameters that can be adjusted by the player both to make changes in tone quality and also to make fine adjustments in tuning. One of the difficulties about all woodwind instruments is that the same tone holes have to be used in different octaves and adjustments in tuning are needed when changing from one octave to another. Also the tone holes are never completely uncovered; even when they are open, the key pads are sufficiently close to the hole to make a slight change in pitch.

4.9 WIND-CAP INSTRUMENTS

Wind-cap instruments have a double reed something like that of an oboe but, instead of the reed being held in the player's lips, it is surrounded by a chamber which is fed with air either directly from the player's lips or via a reservoir as in the bagpipes. The reed is normally open and when air is blown through it the reduction in pressure resulting from the flow leads to the reed closing and the air-flow being stopped. The elasticity of the reed then causes it to open and the cycle is repeated. Thus excitation by a reed consists of a succession of short pulses and the tone is rich in overtones.

There are many variations in this family, including the crumhorn which exists in several pitches and is relatively loud and harsh, the Welsh pibcorn which is a little softer, and the cornemuse which is quieter and has a rather plaintive sound. These all exist today mostly as modern reproductions and are used principally in revivals of medieval bands. The only members of this family that have survived in continuous use are the many varieties of bagpipes. The chanter or melody pipe has a double reed of a similar type to that of the other wind-cap instruments, but the drone pipes, which provide continuous background notes of fixed pitch usually have a single reed; examples of each are shown in figure 4.14.

4.10 MOUTH–REED INSTRUMENTS

The double reed of members of the oboe family is, superficially, similar to

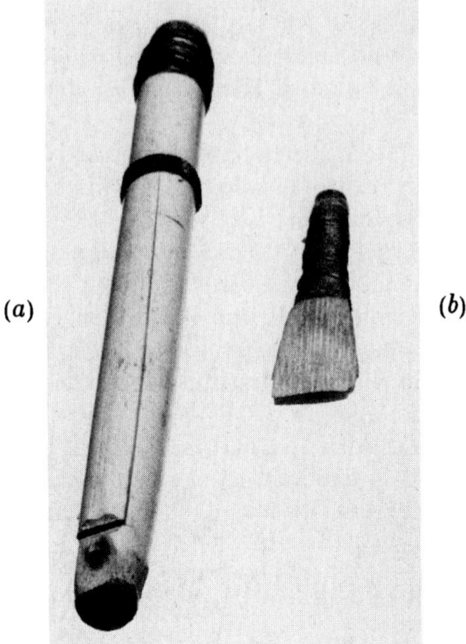

Figure 4.14 (*a*) Single reed of a bagpipe drone, and (*b*) double reed of a bagpipe chanter.

that of a wind-cap instrument but it is held in the mouth with the lips exerting pressure on the reed as an additional control. The oboe itself has a conical bore and hence will produce a full series of harmonics and will overblow at the octave. However, just as with the flute and other woodwind instruments, the side holes and the keywork constitute, in effect, departures from true conical bore and so various tricks both in construction and in playing are needed in order to achieve correct intonation. The length of the oboe is approximately the same as that of a flute and, since the conical pipe behaves like a pipe open at both ends, it will have approximately the same pitch as that of a flute. There are the same six basic tone holes as for the flute and these provide the core scale of the instrument, but additional keys extend the range downwards to just below middle C. The fingering for the upper octaves becomes quite complicated but the top note is approximately 2 1/2 octaves higher.

The alto member of the family is the cor anglais and, apart from its greater size, the main difference is in the bell, which is shaped rather like a Helmholtz resonator. Its main effect on the tone is confined to the lowest two or three notes for which there are no lower finger holes to assist in the radiation of the sound. Because of its greater length the mouthpiece is angled to make it easier for the player to reach the keys.

The tenor member of the family is the bassoon. Here the length is so great that it has to be effectively folded in half.

The clarinet family has a single reed attached to the side of a rigid mouthpiece. The reed rests on the lower lip and, again, the pressure of the lips provides additional control. Clarinets have a cylindrical bore and hence overblow at the third harmonic (an interval of an octave plus a fifth). Thus the basic six or seven tone holes are unable to fill the whole of the gap between the note given with all holes open and that produced by overblowing with holes closed. A relatively complicated system of keys and additional holes with correspondingly more complicated fingering is therefore required than for the flute or oboe. Also, because the tube is cylindrical and virtually closed at one end, the lowest note is an octave lower than for the flute and oboe in spite of the fact that they are all of roughly the same size.

4.11 THE FUNCTIONS OF SIDE HOLES

The early members of the flute, oboe and clarinet families all had simple holes like those of the recorder. But very soon some keys were introduced, particularly one or two keys which, though normally closed, could be opened to provide more positive intermediate notes instead of using half closed holes. A little later keys that were usually open but could be closed at will were used to extend the range of the instrument downwards by a semitone or a tone. Many ingenious systems for pressing more than one key at a time were devised but the real revolution came in the middle of the nineteenth century when Theobald Boehm introduced the system that bears his name, first to the flute and later to other instruments of the woodwind family. His system will be described in section 4.14

Obviously the basic purpose of the side holes is to change the vibrating length of the air column in order to change the pitch of the note. But there are other functions that play almost as important a part in the performance of an instrument: namely controlling the way in which the sound energy is radiated from the instrument, and helping maintain the pattern of vibration by controlling the way in which pulses are reflected from the open end of the instrument. We shall discuss these in turn in relation to the clarinet family.

4.12 HOW THE SOUND GETS OUT

The second of the principal functions of the side holes in woodwind instruments, the first being that they provide a means of selecting the pitch of the note, is that they allow the sound to emerge into the air around the instrument.

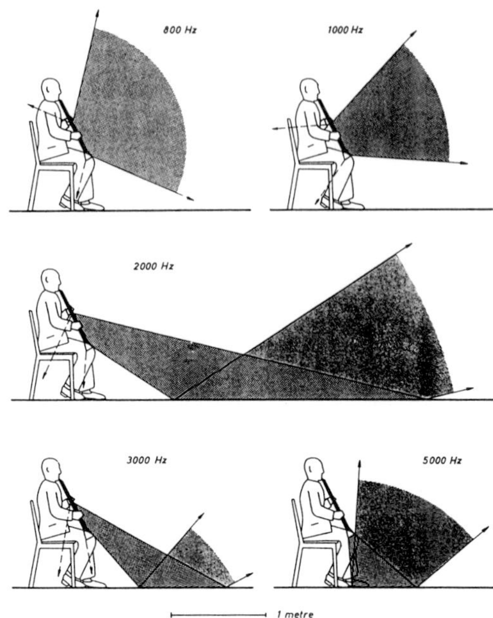

Figure 4.15 The directional radiation pattern for a clarinet at a number of mean frequencies as measured by Jürgen Meyer.

One of my standard demonstrations is to play the note on a clarinet with the three fingers of the left hand all closing holes or keys and the right hand fingers removed. The note produced is then unaffected if the bell is removed, and, still more remarkably, is unaffected if the end of the instrument is completely blocked by the right hand. It is therefore obvious that most of the sound emerges from the finger holes. Although subject to minor deviations the holes are, in effect, a regular array, like, for example, a television aerial. Periodic structures of this sort in physics involving waves always lead to special effects. The reason is simply that there is bound to be some kind of relationship between the spacing of the array and the wavelength. Thus a diffraction grating in optics is a regular array of apertures and the light passing through will be radiated in different directions depending on the relationship of the spacing of the apertures and the wavelength of the light. If white light is used, then different wavelengths (that is different colours) are radiated in different directions. It is not too surprising, therefore, to find that the different frequency components that occur inside a woodwind tube are radiated in different directions.

Jürgen Meyer has made a special study of the directional characteristics of many different instruments. As an example figure 4.15 shows the radiation pattern at a number of mean frequencies for the clarinet. It is interesting in passing to note how the radiation direction influences the studio techniques for recording. If one listens to, for example, a clarinet in a small room the different components are radiated in different directions but the reflections from the walls lead them all to the ears of the listener with scarcely noticeable differences in delay. Because the listener is already in the room and the brain has already, subconsciously, performed an acoustical survey of the distances and directions of the reflecting surfaces, it is able to compensate and the sound of the instrument is acceptable. However, if a recording is made, for example with a directional microphone in a relatively dead studio there can be considerable differences in recorded level for different frequencies that are due entirely to the directional effects. Also when the recording is played back the listener has no means of assessing the directional properties of the room.

4.13 KEEPING THE VIBRATIONS GOING

An experiment with a woodwind instrument and a small microphone that can be inserted into a side hole soon shows that the oscillations are not confined to the section of the tube between the mouthpiece and the uppermost open finger hole. Such an experiment with a microphone placed near the bell shows that the amount of energy emerging from the bell is small and also that it is only high frequency components that are being radiated. Why then do woodwinds have a bell at all? The primary reason is that, for the lower two or three notes there are no finger holes from which the sound can be radiated, and this would lead to a marked change in sound quality which is undesirable. Ideally a short additional section of tube with side holes, designed for radiation rather than for playing additional notes might be a solution. But it would look rather odd and the bell, if properly designed, fulfils the same function.

Benade has shown that the recipe inside the tube is not necessarily the same as that radiated from the side holes. We met a similar effect in talking about the violin. In section 3.8 an experiment was described in which the waveform of the vibrations of the bridge, of the body and of the radiated sound were shown to be quite different. The body, because of its non-uniform amplification properties, imposes a formant characteristic on the waves being generated; the pattern of side holes imposes a similar formant on the waves inside the clarinet.

A demonstration that shows the important part played by the finger holes in determining the oscillation patterns in a clarinet uses three pieces of plastic tubing as shown in figure 4.16. The first (a) is a piece of plain tubing, about 20 cm long, with a clarinet mouthpiece attached to one end. It will play but the note is rather muffled. The second (b) is a longer piece of tubing (about 30 cm) with five holes in the side so that the note produced is at about the same pitch as for (a). Now the sound is much

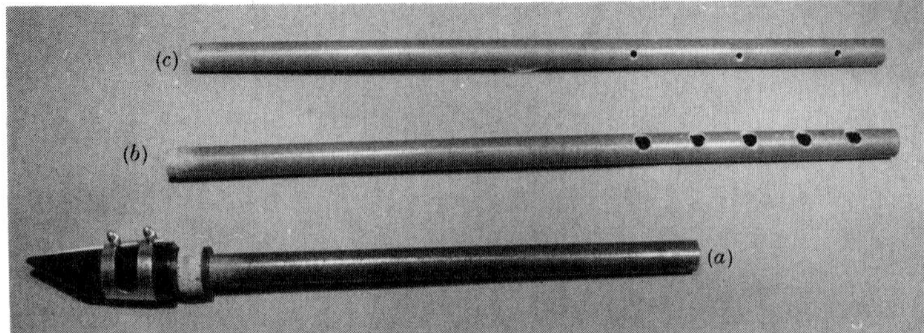

Figure 4.16 Three plastic tubes used to demonstrate the influence of the tone holes on the quality of the note produced by a clarinet-like instrument. In (*a*) the tube has no side holes and the sound is rather muffled. In (*b*) there are five large holes which define an oscillating length roughly the same as that for (*a*). The note is of approximately the same pitch but is much louder and more 'clarinet-like'. In (*c*) there are only three holes and they are much smaller. The tube is very difficult to play when all the holes are open. If two or more are closed a muffled note can be obtained but dies away as soon as all the holes are opened.

brighter and more 'clarinet-like'. The five holes are radiating the sound much more effectively than the plain open end of (*a*).

The third tube (*c*) is the same length as (*b*) but has only three side holes and they are much smaller. This is almost impossible to play at all! But if two or three of the side holes are closed a note can be produced. In this case the side holes are located at exactly the wrong positions for maintaining the oscillations.

It is not my intention to delve deeply into the theory of this demonstration; my purpose is simply to emphasise the way in which real instruments differ from simple tubes and to pay tribute to the skills of instrument makers in coping with all these complexities. But the improvement in our knowledge of the physics of clarinets—largely due to Professor Benade—is too important and interesting to miss out altogether. So I plan to include a greatly simplified, but, I hope, fair description that will at least give a flavour of the most exciting and elegant ideas that have emerged over the last twenty five years or so.

We have already likened the regular spacing of the side holes on a woodwind to the elements of a television aerial or of an optical diffraction grating (section 4.12) in relation to radiation. But it also turns out that, just as one needs to alter the size and spacing of the rods on an aerial for different frequency bands, so the size and spacing of the holes has a bearing on the response frequencies of the pipe. If the initial wave travels down a plain section of pipe with no holes and then meets a section in which there are regularly spaced holes, the lower frequency components of the internal

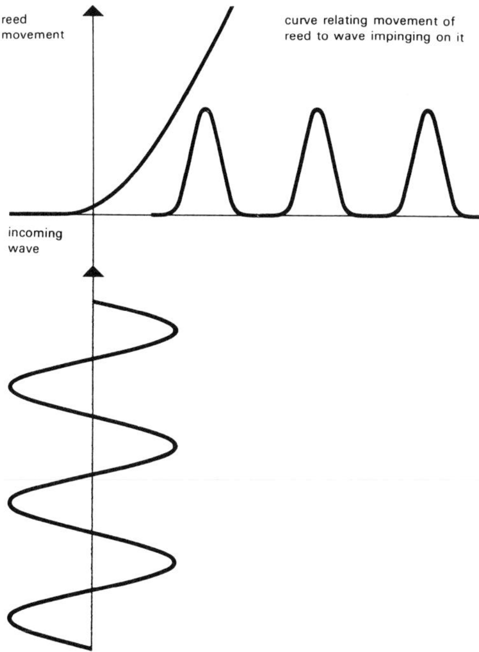

Figure 4.17 Graph showing how a sinusoidal input (coming from below) can be converted into a pulse sequence (emerging to the right) as a result of non-linearity in the reed.

recipe will be reflected from the first open hole just as simple theory would predict for an open end. But higher frequency components 'leak' past this hole and travel down to the end, ignoring the open side holes. Careful investigation shows that there is a specific frequency at which these two kinds of behaviour change over. This frequency is known as the cut-off frequency. Its value depends on the sizes and spacings of the holes in relation to the bore diameter of the pipe and also on the wall thickness (which dictates the length of the path through the hole between the inside of the tube and the outside air). The bells of woodwinds and the horns of brass instruments also introduce cut-off frequencies which affect both the recipe of the internal oscillations and of the radiated sound.

Now we come to yet another complication! When one delves into the real facts about the behaviour of pipes which are not plain cylinders or cones, but have side holes, or bumps in the bore resulting from closed side holes, it turns out that the overtones are not strictly harmonic. This might be expected to alter the sound quality but perhaps not to have any more major consequences. But in fact because of the non-linear behaviour of the reed they turn out to have considerable consequences. What exactly do we mean by 'the non-linear behaviour of the reed'? It simply means that the reed does not vibrate in the manner that a physicist would call 'sinusoidal'. A pendulum oscillates sinusoidally and makes symmetrical movements on either side of the midpoint. But by its very nature a reed cannot behave symmetrically on either side of its rest position. For example, a clarinet reed is much more severely restricted in its movement towards the edges of the mouthpiece than in its movement away. Figure 4.17 shows how a wave reflected back and moving towards the reed from the body of the instrument is translated into a pulse sequence by the reed non-linearity. This will happen for all the frequency components of the reflected waves, but only if they are harmonically related to each other will the reed be able to maintain all of them. If they are not harmonically related, one or other of the components will predominate and the others will get out of step. The resulting lack of cooperation will certainly diminish the amplitude of vibration and may even stop it altogether. This is a partial explanation of the behaviour of the third of the plastic tubes in the demonstration illustrated in figure 4.16

So we can see that there are three distinct functions of the finger holes. First to determine the pitch of any given note; secondly to control the maintenance and recipe of the oscillations in the tube; and thirdly to control the radiation of sound. But these three cannot be adjusted independently unless there are parameters other than the position of the holes that can be changed. These additional parameters are the diameters of the holes, the wall thickness, and the internal bore. Diameters of holes are easy to change and have a significant effect on the pitch of a note. Figure 4.18 shows a set of identical descant recorder mouthpieces fitted to five plastic tubes of the same internal bore. Four of the five have a side hole at the same position but of varying diameters and the fifth is cut short at the position of the hole in the others. The frequency produced by gentle blowing is shown in the caption.

In a clarinet or oboe the wall thickness can be changed either by counterboring a larger hole around the actual finger hole or by raising a wall round the hole. Examples of each method, and of variations in hole size can be seen in figures 4.19 and 4.20. The bore can be changed, as already mentioned, by special reamers that can be made to expand at a particular point or by painting on lacquer at particular places using special brushes. Changes in the bore measured in microns can be significant. The process is easy for any one note. But the problem of adjusting each note without upsetting all the rest is a major one.

Professor Benade developed the art of encouraging cooperation between the modes in a clarinet for each individual note to such a point that he could make exciting improvements in both tone, and in ease of playing, in already good instruments. But be warned, his process involved immense

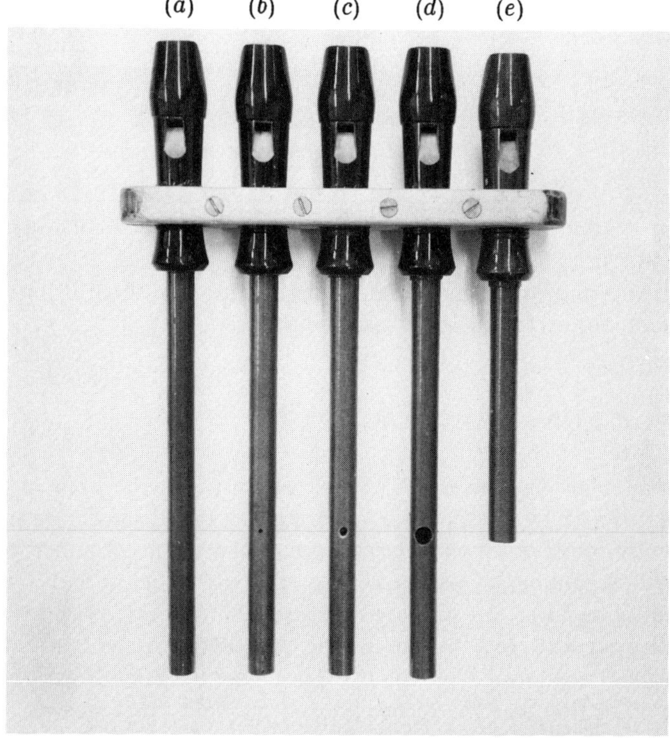

Figure 4.18 Set of five identical recorder heads attached to plastic tubes: (*a*) has no finger holes; (*b*), (*c*), and (*d*) have holes placed at the same position but of different sizes; (*e*) is cut off at the position of the holes on the other tubes. A simplistic view might suggest that tubes (*b*), (*c*), (*d*), and (*e*) should all give approximately the same note. In practice all five give different frequencies which are: 538, 587, 622, 668, and 698 Hz, respectively.

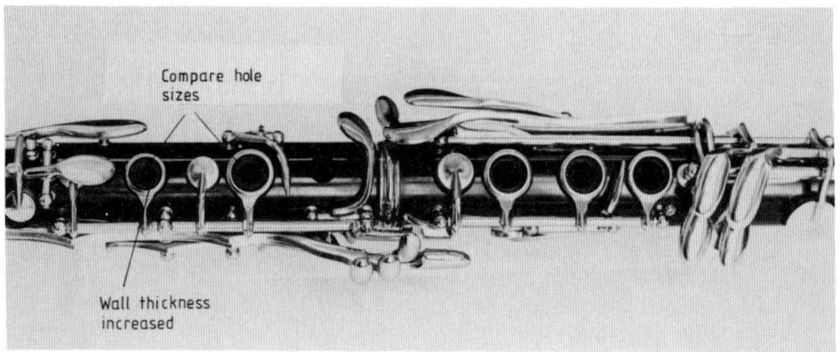

Figure 4.19 Close up view of the keys and finger holes of a clarinet.

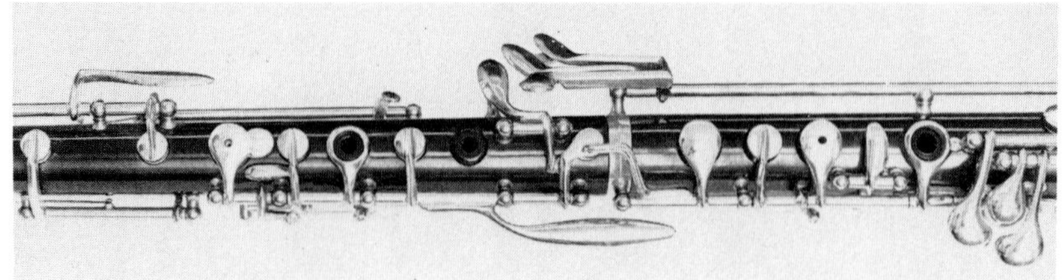

Figure 4.20 Close up view of the keys and finger holes of an oboe.

skill, patience and experience, and required continual monitoring of input impedances (see section 4.15) at all frequencies.

4.14 THE FUNCTIONS OF KEYS

Boehm was a jeweller by trade and so was used to the kind of detailed metalwork that now characterises instruments based on his designs. He was also a flute player of considerable competence who occupied principal positions in a number of professional orchestras. Boehm had a scientific turn of mind and was not prepared to accept all the myths and traditions that tended to surround the making of musical instruments at that time and to bedevil any new developments. He concerned himself with two main problems. The first was the rather poor tone quality that was usually produced by fork or cross fingering of semi tones. And the second was that

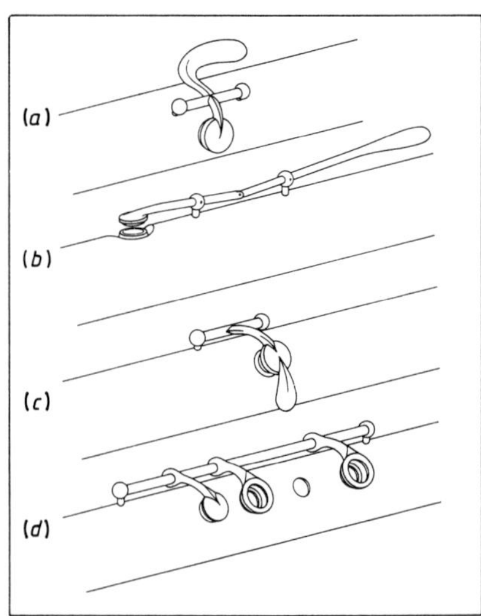

Figure 4.21 Diagrams of four different types of woodwind key: (*a*) closed key; (*b*) open key; (*c*) hinged key; (*d*) ring key.

flute makers at that time used relatively small side holes, placed them in the most comfortable position for the player and then changed the diameters to bring the notes into tune. The result was non-uniformity of tone. Boehm realised that enlarging the holes would produce a louder tone and that placing them in the musically desirable positions could improve the uniformity of tone. However, such a flute would then be unplayable with any normal human hand. Hence his development of the key and fingering system that now bears his name.

Keys perform four main functions on woodwinds:

(i) they permit holes to be covered or uncovered that cannot otherwise be reached by the fingers;
(ii) they enable greater numbers of holes to be controlled than is possible with the normal human complement of fingers;
(iii) they enable larger holes to be covered than with normal fingers;
(iv) they permit the performance of rapid passages or trills that would otherwise involve such complex changes of fingering as to make fluent execution impossible.

There are four basic types of key used on instruments of the woodwind family and typical specimens are shown in figure 4.21 and are described below:

(a) closed keys in which pressure lifts the pad from the hole;
(b) open keys in which pressure lowers the pad onto the hole to close it, and which can be operated at a point some distance away from the hole;
(c) hinged keys, which are a version of the open keys but operate near the site of the hole and whose main purpose is to close a larger hole than could be closed by a finger alone; sometimes such a key is pierced with a small hole, so that by moving the finger a little to one side an intermediate pitch can be obtained;
(d) ring keys, which are an ingenious method of controlling more than one note with the same finger.

> Suppose, for example, that there are four finger holes to be controlled by only three fingers. Such an arrangement on a clarinet can be seen in figure 4.19. The highest hole is fitted with an open key (type (b) above) which can be closed by depressing any one of the three ring keys on the next three lower holes. The fingers remain over these three ring keys throughout. If all three fingers are down then all four holes are closed. If finger 3, or fingers 2 and 3 are raised the corresponding holes are opened but finger 1 keeps the open key on the highest hole closed. If finger 3 is depressed and finger 1 raised, the open key is kept closed by finger 3, which is far enough away not to influence the

pitch. Of course if all three fingers are raised then all four holes are open.

Many variations and combinations occur and makers tend to make modifications to suit the needs of individual players.

The Boehm system and various modifications have had a profound effect on the composition of music for the woodwinds. The possibility of controlling more holes with far less movement of the fingers from their basic positions, the provision of alternative keys to play the same note in different sequences, and the provision of alternative keys to enable trills to be played that would normally involve the thumb (which is needed for supporting the instrument) have all added greatly to the fluency and speed of playing.

One final point should be mentioned here although it is not so directly concerned with keys. As instruments become larger the separation between finger holes may become too great for the human hand. The bassoon (see section 4.7 above) is a case in point. Bassoon makers have solved the problem by adding a thick wall (called the wing) to the instrument and then drilling the finger holes at an angle. Thus the holes are close enough at the outside of the instrument to be reached by the fingers and yet at the point where they enter the main bore are far enough apart to suit the tuning.

There are other possible ways of overcoming the difficulty of hole spacing, but it turns out that there is a marked effect on the tone quality caused by the extra length of the oblique holes and this has now become an accepted part of bassoon quality.

4.15 ANOTHER VIEW OF VIBRATIONS IN TUBES

A major advance in understanding the behaviour of wind instruments in general was made when experimental methods of measuring the input impedance of instruments were devised. Acoustic impedance is analogous to electrical impedance. For an electrical circuit the impedance relates the applied voltage to the resultant current. Thus a low impedance will allow a high current even with a very low applied voltage. The electrical impedance becomes the same as the resistance if direct current is involved. But if alternating current is in use then the impedance becomes a mathematically complex quantity and not only relates the size of the voltage to that of the current, but also the relative phases of the two, and becomes dependent on the frequency of alternation.

In the acoustic case the relationship is between the applied pressure variation and the resultant air-flow variation. It is also a complex quantity and involves phase relationships as well as amplitudes. It is also frequency sensitive.

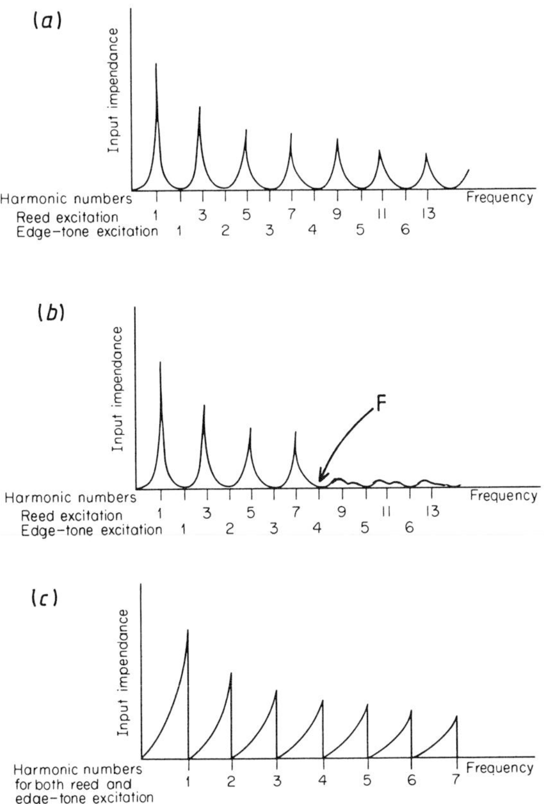

Figure 4.22 (*a*) Curve showing the relationship between input impedance and frequency for a cylindrical tube. (*b*), As (*a*) but for a tube with a series of open side holes giving rise to a cut-off frequency at F. (*c*) Curve showing the relationship between input impedance and frequency for a conical tube.

As examples the edge-tone instruments are low impedance devices; the actual pressure alternations are quite small but the oscillatory flow is large. On the other hand the reeds are high impedance devices in which a high pressure alternation results in a relatively small oscillatory flow. Watch an oboist playing an extended passage; at the point at which a singer, or a flautist would take a breath, the oboist has to breathe *out* in order to relieve the pressure. It is quite easy to measure the input impedance, for example, of a tube at one specific frequency; the problem arises when a rapid method of measuring it over a wide range of frequencies is required. The apparatus is very similar to that of figure 2.11. An alternating air-flow is applied by means of a loudspeaker unit which is fed by a variable frequency oscillator. A pressure microphone measures the pressure, but its electrical output must be filtered so that only the pressure alternations at the specific frequency being applied are measured. It is thus necessary to use a tunable filter whose frequency is always the same as that of the driving oscillator. The whole system is usually automated so that the frequency of both oscillator and filter is swept over the whole of the required range and the ratio of pressure to flow is plotted automatically.

Figure 4.22(*a*) shows a typical plot for a simple cylindrical tube. Immediately one can see that a high impedance device like a reed will match satisfactorily only at the points marked 'Reed excitation', 1, 3, 5, etc. In other words a tube like this used as a simple clarinet will produce only the odd harmonics of the fundamental.

On the other hand, edge-tone excitation, which is necessarily low impedance, will work only at the points marked 'Edge-tone excitation', 1, 2, 3, 4, etc. We can also see that the fundamental for low-impedance excitation is an octave higher than for the reed and the full harmonic series is produced. In section 4.13 we talked about the cut-off frequency. In figure 4.22(*b*) we see the input impedance curve for a tube with a series of side holes like those of a clarinet for comparison with the plain tube. The cut-off frequency is clearly shown at F.

If the input impedance measurement is made for a conical tube the symmetry of the peaks is upset and the general appearance is like that of figure 4.22(*c*). Now the highest and lowest impedance points are extremely close together and, as we saw in section 4.6, the result is virtually the same whether high- or low-impedance excitation is used and the tube gives a full series of harmonics.

Apart from confirming the general behaviour of woodwind instruments, the input impedance approach helps to make much more sense of the behaviour of brass instruments than could be achieved with the more simplistic approach of section 4.6. Indeed it is principally for that reason that the approach has been introduced. However, before considering the modern Brass family it will be useful to consider what might be referred to as 'transition instruments' in that they are excited by vibrations of the lips, as with modern brass, but have finger holes or keys like woodwinds to fill in the gaps between harmonics in order to play tunes.

4.16 TRANSITION INSTRUMENTS

At the end of section 2.9, mention was made of the tabor pipe in which, although an edge-tone instrument, overblowing to quite high harmonics was used in order to reduce the number of side holes required so that tunes could be played with one hand while the tabor was beaten with the other. However, instruments like the cornett and serpent were true lip-reed instruments like the modern brass, but had finger holes to permit the wide gaps between the lower harmonics to be filled in order to play tunes.

Figure 4.23 Playing a reproduction serpent. On the bench are some examples of modern brass instruments and, just in front of the trombone, is a treble recorder and a cornett, both modern reproductions.

Both instruments were originally made of wood. The instruments were made in two halves, split along the length. They were glued together and often covered in leather; examples of each can be seen in figure 4.23. Modern reproduction instruments are made in synthetic materials. The high pitched cornett was at the height of its use in the sixteenth and seventeenth centuries, for example by composers like Monteverdi. The much lower pitched serpent appeared a little later but persisted in church bands until the middle of the nineteenth century.

Both instruments originally had six finger holes and latterly often had one or two keyed holes to extend the range. The shape of the serpent was largely determined by the need to reach the holes with the fingers. A typical serpent would be about 8 feet long if straightened out. It will be immediately obvious that, even with the serpent-like shape, the finger holes cannot possibly be both easily accessible to the fingers and in the musically correct positions. However, the nature of these instruments is that the dominant factor in determining the pitch of the note is the lips of the player. One of the most important requirements of a player, therefore, was a good 'musical ear'. The note could be pulled into tune even if the positions of the finger holes were quite wrong.

The serpent was replaced orchestrally by the ophicleides. These are made of brass and have cup mouthpieces like modern brass instruments, but they have a conical bore and have finger holes which are mostly fitted with keys. They existed in many different sizes, those of the highest pitch were really keyed bugles. The lower pitched members of the family appear in the early compositions of composers like Mendelssohn, Verdi and Wagner. The fundamental problem with the ophicleide and key bugle family is that it is virtually impossible to find side-hole positions that are a sufficiently good compromise to enable notes to be played in tune while retaining some semblance of constant tone quality. Hole positions set for the fundamental mode would be too far away from the ideal positions for the second and third modes so that, even though it may be possible to 'lip' the notes into tune, the tone quality will be completely different. The ophicleides have now been completely replaced by the tubas, which we shall discuss towards the end of the next section.

The forbears of the modern brass family were the bugle and the post horn. Each of these is capable of playing only in a series of modes so their tune-playing capabilities are severely restricted.

4.17 TURNING A TUBE INTO A TRUMPET

When the average young physicist tries to relate the elementary ideas learned from experiments on the resonance tube in a physics lab with the behaviour of a real trumpet difficulties always arise. Take a straight length of brass tubing with an internal bore of about $1\frac{1}{2}$ cm and about $1\frac{1}{4}$ m long. Wrap some drafting tape round the end to thicken it a little and hence make it more comfortable when the lips are applied. The approximate notes you will get using the lips as reed to excite vibrations will be:

$$204 \text{ Hz}, 340 \text{ Hz}, 476 \text{ Hz}$$

and you will recognise these as the third, fifth, and seventh harmonics of 68 Hz. This series of odd-numbered harmonics is the one you would expect to get from a tube closed at one end. The wavelength to be expected from a tube of this length closed at one end is 4×1.25 m $= 5$ m, and assuming a velocity of sound of 340 m per second agrees with our figure of 68 Hz for the fundamental. You are unlikely to be able to produce the actual fundamental note.

You will immediately realise that these are not the notes that you would get from a bugle or trumpet, and, furthermore, the quality of the note is very far from the normal brilliance of a trumpet. It is quite extraordinary to realise that the conversion of the simple tube into the modern trumpet was done over a long period of evolution, by totally empirical methods with no real understanding of the physics of the process! Success was entirely due to the patience, persistence and skill of generations of instrument makers. With the hindsight gained through relatively recent electronic techniques of impedance measurement we can now piece together how, in fact, the instrument makers actually achieved the result.

In order to improve the comfort of the lips a properly shaped mouthpiece is used. The word 'properly' conceals a remarkable evolutionary process. It turns out that the mouthpiece behaves differently at different frequencies. At the lower end of the range the mouthpiece merely behaves as though a piece of tubing of the same bore as the rest of the tube has been added on, of such a length that it has the same volume as that of the mouthpiece. It further emerges that the mouthpiece behaves like a Helmholtz resonator, and indeed the natural frequency of this resonator can be found by slapping the large end of the mouthpiece with the palm of the hand. Above this natural frequency the effect of the mouthpiece is to add on an additional length of pipe and the equivalent addition increases as the pitch rises.

The remarkable fact is that this addition lowers the pitches of the higher harmonics by just the right amount to convert the sequence of odd harmonics into a full sequence!

The lower notes, so far unaffected by the addition of the mouthpiece, now need to be raised by successively increasing amounts as we move to lower pitches. This is accomplished by flaring the end of the trumpet to give the familiar bell. By a subtle combination of adjustments to the bell shape and to the mouthpiece shape, and by adjusting the final overall length the full series of harmonics that one would expect from a tube open at *both* ends is achieved (except for the fundamental, which is out of tune, but, since it is never called for, this does not matter). And this, of course, is the source of the mystification experienced by physicists that was mentioned earlier!

The addition of the flare has another profound effect on the trumpet. The whole basis of the operation is that energy is constantly fed in at the right frequency by the player, but a proportion is reflected back from the open end in order to define the frequency, and a proportion is radiated from it. The precise shape of the flare controls these proportions and they vary over the whole frequency range. The flare enhances the radiation from the end of the tube, particularly at the higher frequencies and so controls the 'recipe' of the various harmonic components that are generated by the tube from the raw pulses fed in by the lips. The behaviour of horns was touched on in section 4.13 in which we said that a horn can produce a cut-off frequency which determines the frequency below which most of the energy is reflected and above which some energy leaks out. Two experiments can be done to illustrate this.

> A 3 m length of garden hose will produce a sequence of notes if excited by the lips, but an ordinary funnel (such as is used in pouring oil or other liquids from one vessel to another) attached to the end will greatly increase the volume of the sound and change its quality. And a mute placed in the end of a trumpet modifies the quality and volume of the sound by altering the

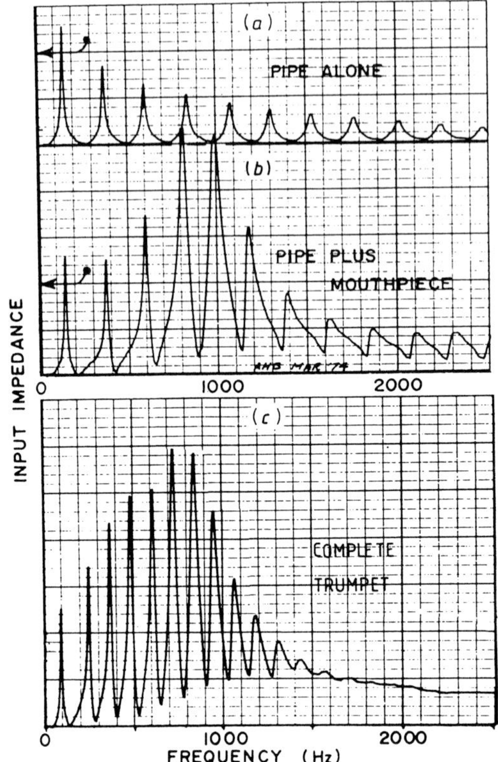

Figure 4.24 Input impedance curves produced by Art Benade for (*a*) trumpet pipe alone; (*b*) trumpet pipe with mouthpiece added; and (*c*) complete trumpet.

proportions of sound reflected and radiated at various frequencies. The French horn player keeps one hand inside the bell of the instrument and by varying its position can change the cut-off frequency, and hence the quality of the sound.

The very significant influence both of the mouthpiece and of the flare are illustrated in figure 4.24 which shows how the input impedance curves change as the additions are made.

4.18 VALVES AND SLIDES

The early trumpets and horns were capable of producing only notes which belonged to a harmonic sequence. Such a sequence written out in normal musical notation for a modern trumpet in B$^\flat$ is shown in figure 4.25. These would be the only notes possible if the trumpet had no valves, as was the

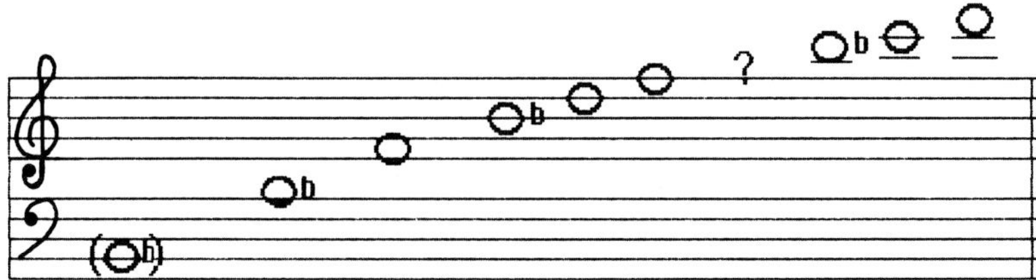

Figure 4.25 Sequence of possible notes for a 'natural' B$^\flat$ trumpet, that is one with no valves. The fundamental, or pedal note, which is never used because it is out of tune, is shown bracketed.

case for so called 'natural' trumpets and it is immediately obvious that if it is desired to play tunes other than simple bugle calls, the *lowest* note would be $B^\flat_5$, which is almost two octaves above middle C. In other words only tunes at quite a high pitch would be playable. Also it would be possible to play only in the key of B flat, or whatever was the natural key of the trumpet. At the time of Bach these were the only kinds of trumpets available and, if you listen, for example to a piece such as the first movement Brandenburg Concerto No 2, in F, you will find that whenever the tune is passed from, say, the recorder or flute, to the trumpet the whole tune is raised much higher.

The problem of playing in other keys could be solved by the addition of 'crooks'. A section of the trumpet tube was removable and could be replaced by sections of different lengths. Of course this took time and could be done only when the player had a significant rest.

The third way of 'filling in the gaps' between the harmonics (the first was the use of side holes and the second the use of crooks) is to use a sliding crook, as for example in the trombone. With this the player can lengthen the vibrating length of the tube by a very considerable amount. The big advantage is that the player can adjust the length to be exactly right for the particular mode in use and no compromise is needed. The only problem is the technological one of making the air-tight slide which will, nevertheless, move very freely, which presents quite a challenge to the instrument maker.

The fourth solution, which is used on all the modern brass instruments except the trombone, is to fit three or four crooks which can be brought into use as needed by using piston or rotary valves. On a three-valve system, such as is commonly used in trumpets, depression of valve one effectively extends the length enough to lower the pitch by a whole tone; depression of valve two lowers the pitch by a semitone; and depression of valve three lowers it by one and a half tones.

Thus, on the B$^\flat$ trumpet already considered, the biggest gap (the lowest note is never used musically) is that between the second and third harmonics, that is between $B_3^\flat$ and F_4. The table below shows how this gap would be filled.

Table 4.1 Showing how the gap between harmonics 2 and 3 on a $B^\flat$ trumpet are filled.

Note required	Valves depressed	Total lowering of pitch
F_4	–	0
E_4	2	$\frac{1}{2}$ tone
$D_4^\#/E_4^\flat$	1	1 tone
D_4	3	$1\frac{1}{2}$ tones
$C_4^\#/D_4^\flat$	2 + 3	2 tones
C_4	1 + 3	$2\frac{1}{2}$ tones
B_3	1 + 2 + 3	3 tones

I said a little earlier that the slide of a trombone can adjust the sounding length of the instrument to be exactly right without compromise. But valves cannot be exactly right for all notes. In section 1.5 we discussed the relationships between pitch and frequency and from that discussion it can be seen that, to change any given note by a particular interval we have to make a *percentage* change in the frequency. Thus the interval of a musical fifth (frequency ratio 3/2) above any note involves increasing its frequency by 50%. Thus a fifth above A_4 (440 Hz) is E_5 (660 Hz) and a fifth above E_4 (330 Hz) is B_4 (495 Hz).

The middle valve (2) on a trumpet lowers the pitch by a major tone (frequency ratio 9/8), which needs an increase in length of 12.5%. Thus if we adopt for the sake of argument a total tube length of 100 cm, the addition required when the middle valve is pressed is 12.5 cm. But, if the pitch has already been lowered 1 1/2 semitones by valve 3, its effective length will be 118.75 cm and to lower the pitch a further tone using valve 2 we should need to add 12.5% of 118.75 which is almost 15 cm. But we have already seen that valve 2 only adds 12.5 cm to the length, and hence the note would be about 1/3 semitone sharp.

With the higher-pitched instruments like the trumpet a skilled player can easily dominate the instrument and 'lip' the note into tune. But with the very large brass instruments like the tuba this is more difficult and ingenious compensation systems have been developed.

A little earlier in this section we said that trumpets in Bach's day were capable of playing tunes only in the upper registers. One of the immediate consequences of the development of valves was that composers could then write pieces for the trumpet that included notes in all registers.

In section 4.5 the existence of privileged frequencies was discussed. There are so many of these that they assist greatly in enabling skilled players to glide smoothly from one note to another even on a valved instrument. The beautiful demonstration of the use of privileged frequencies on the trombone by Dr John Bowsher, was also described in the same section.

4.19 HARMONIC RECIPES IN THE BRASS FAMILY

A striking way of demonstrating the differences in tone quality which arise largely from differences in the harmonic recipes, is to listen to the sound produced by four instruments of approximately the same length. For example a euphonium, a baritone horn, a tenor trombone, and a piece of plain brass tubing a little over 8 foot long all have approximately the same length. But even if played with the same mouthpiece, the tone quality is quite obviously different. Figure 4.26(a)–(d) show the differences in the wave traces and figure 4.27(a)–(d) show respective harmonic analyses when each instrument sounds the same note: $B^{\flat}_2$, 116.5 Hz.

The differences in quality and recipe are obvious. The euphonium is the most mellow and has rather fewer high harmonics; the tenor horn is rather brighter in tone and has a clearly different harmonic recipe. The trombone is the most significantly different and, in particular has higher harmonics that are even stronger than the fundamental. It must be stressed that, just as in section 4.7 with the woodwinds, the quality of all the instruments varies over the pitch range. The simple tube has a very muffled quality.

The origins of all these differences lies in the different shapes. The trombone is cylindrical for a large part of its length. This feature is, of course, necessary to enable the slide to work. the flare towards the bell is also relatively slow. The baritone horn has cylindrical portions, but a large part of it is more or less conical and the flare to the bell occurs over a longer portion than for the trombone. The euphonium has a much fatter bore throughout and the final flare to the bell is much more conical

The design of all these instruments is an immensely skilled process and relating the resultant quality to the parameters of the shape is not an easy scientific process!

4.20 ORGAN PIPES

The organ is perhaps the greatest of all the wind instruments since it really brings a complete orchestra of instruments under the direct control of a single player. The organ belongs to the first family of instruments (see section 1.9) in that it is really an enormous collection of instruments each of which is capable of producing only a single note.

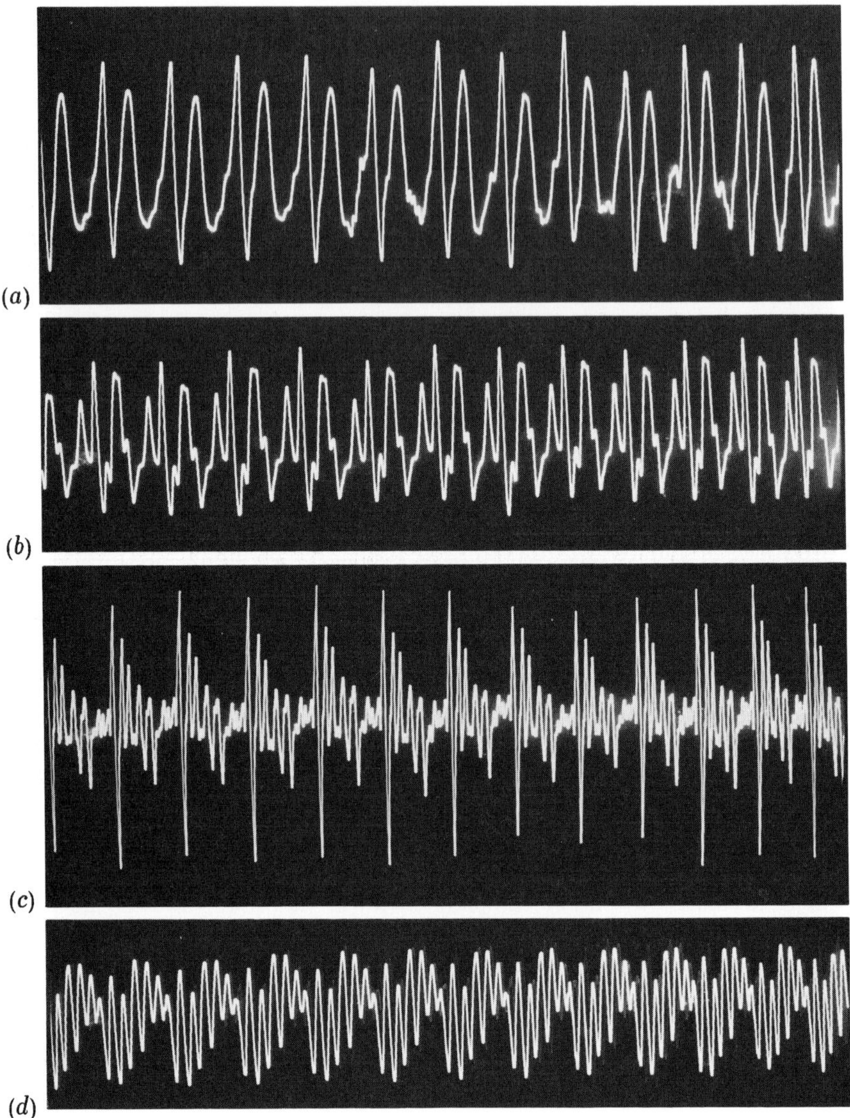

Figure 4.26 Oscilloscope traces of the waveforms of the note $B^\flat_2$ (116.5 Hz) on: (*a*) a euphonium; (*b*) a baritone horn; (*c*) a tenor trombone; (*d*) an 8 foot length of brass tubing.

The basic principle of operation is the same whatever the size of the organ, though the technical means of achieving the various functions may vary. All the pipes are grouped in three possible ways. First, all the pipes of whatever kind but of the same pitch are linked together and controlled by a particular key (which may be on a manual or a pedal keyboard); secondly

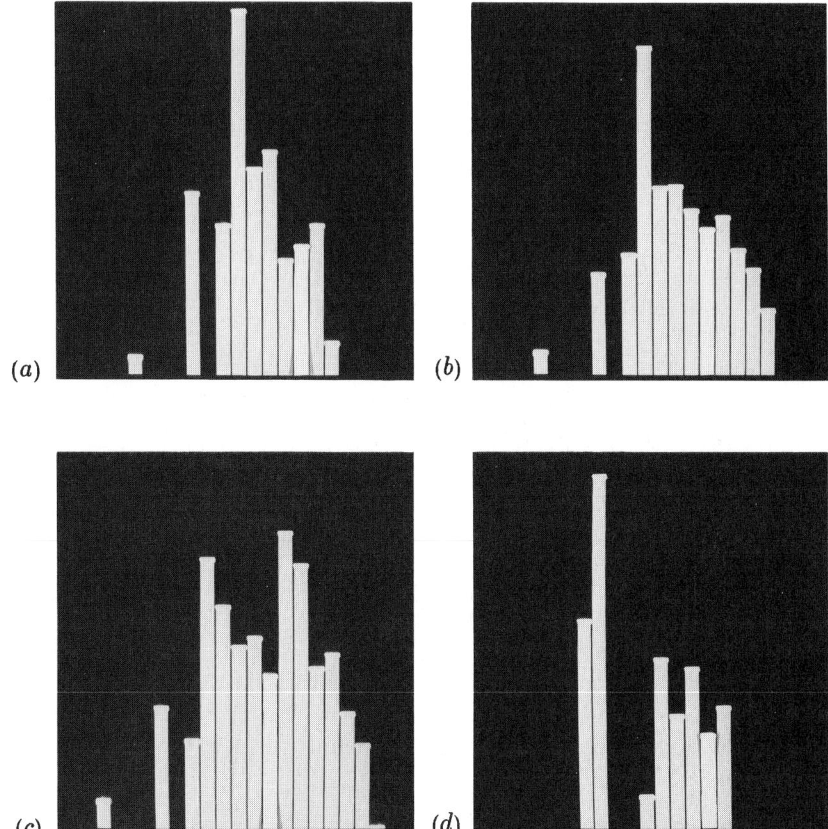

Figure 4.27 Frequency analyses of the notes corresponding to figure 4.26.

all the pipes of the same quality, whatever their pitch (the collection is called a 'rank' of pipes) are linked together and controlled by a given stop; and finally groups of related keys (e.g., sub-octaves, octaves, etc, on the same keyboard, or notes of the same pitch on different keyboards) or stops in given combinations are linked by couplers. On a large organ a common chord of four notes played on one manual could easily result in several hundred pipes of differing tone qualities, and pitches, normally controlled from different keyboards, to sound simultaneously. The actual mechanism of control and other technical details will be described in the next section.

There are two basic types of pipe found in all organs, though the variations in detail and tone quality in each are legion. Edge-tone excited pipes are known as 'flue' pipes; the reeds, as their name suggests involve reed excitation. The flue pipes (see figures 4.28(*a*) and (*b*)) are very much like recorders but without any finger holes and, of course, can produce only one note. There are two main categories; pipes open at the upper end, which

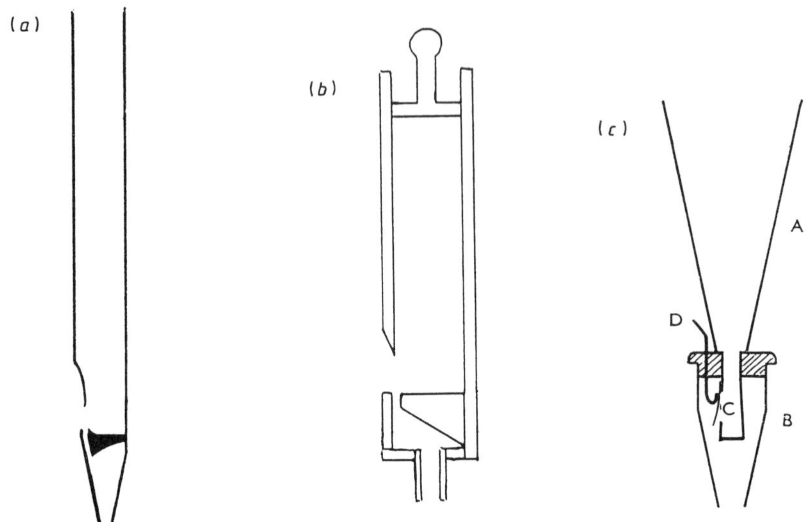

Figure 4.28 Some typical organ pipes: (*a*) vertical section through a metal open diapason pipe of circular cross-section. Tuning is by opening out or closing in the upper end of the pipe with special conical tools. (*b*) Vertical section through a stopped diapason wooden pipe. This pipe is square in cross-section and tuning is by moving the stopper in and out. (*c*) Section through a metal reed pipe: A is the body of the pipe; B is the boot containing the reed, C; D is the rod for tuning the reed.

include the open diapasons that provide the foundation of organ tone and the stopped diapasons, which give a more flute-like tone. Both can be made of metal and are then usually cylindrical, or they can be of wood, usually of square section. The speaking length can be as small as 1 cm (which for an open diapason gives a note of around 15,000 Hz) or as long as 5.5 m (giving a note of about 30 Hz). The pitch (see section 1.5) of a rank of pipes is usually designated by the length of the pipe giving the lowest note. The normal pitch (in which middle C is 260.7 Hz) is conventionally described as '8 foot pitch' because the lowest note on a standard organ manual (which is two octaves below middle C and designated C_2) has a frequency of 65.2 Hz, which would be sounded by an open pipe 8 foot long (giving a wavelength of 16 foot). Ranks of pipes with this pitch, or that are octaves or sub-octaves of it (e.g., 16 foot, 4 foot, 2 foot, etc), are known as 'foundation stops'. Pipes that sound other harmonics or sub-harmonics of this (e.g., 2 2/3 foot pitch, which sounds the third harmonic of the normal pitch) are known as mutation stops.

The open diapasons, as their name suggests are open at the upper end and so produce a full series of harmonics. Tuning can be performed in several ways. The metal of which organ pipes are made is relatively soft and the smaller cylindrical pipes can be slightly opened out or closed up by means of special conical tools. The larger pipes may have a tongue cut out at the top which can be bent slightly inward or outward. Wooden pipes may be fitted with a metal tongue that can be bent as with a metal pipe or may have a short sliding section that can be moved up or down.

The stopped diapasons or flutes have a stopper in the end and so give only the odd harmonics. They also tend to have a slightly more 'breathy' sound probably because it takes a little longer for the musical tone to build up and so the sound of the initial onset of air-flow into the pipe is not masked as much as for the open pipes. Tuning is simply by moving the stopper in and out.

Enormous variations in tone can occur even with this basic pipe. The variables that can affect the quality are the internal bore, the size of the slit through which the air is delivered, the wind pressure applied, the angle of the wedge at which the air is directed, the height of the wedge above the slit, the position of the wedge in the air stream from the slit, the addition of 'wings' on either side of the wedge and many others. The building of organ pipes and their subsequent 'voicing' when the organ has been assembled are enormously skilled operations. Most organ builders operate on a basis of 'rules of thumb' acquired through many years of experience and are often unaware of any but the most basic physical principles involved in the operation. Furthermore, a physicist who is expert in musical acoustics can find fascination, much food for thought and considerable problems in watching a skilled organ builder at work. Trying to provide physical explanations of how a particular technique for modifying the tone quality of a pipe actually achieves its end is often extremely difficult!

The orchestral instruments most closely resembling organ reeds are the wind-cap instruments (section 4.9). The reed itself, which is usually of springy brass or bronze, is attached to a metal support at the foot of the pipe and hangs in a wind chamber usually called the boot (see figure 4.28(*c*)). The pipe itself occurs in an enormous variety of shapes, mostly conical. Often the tops of the reed pipes are bent to face the audience since most of the radiated sound emerges from the end. Tuning involves two adjustments that are not entirely independent of each other. The rod D in figure 4.28(*c*) is used to adjust the vibrating length of the reed and then the effective length of the pipe is altered to bring it into resonance.

The fact that organs belong to the first family with separate instruments for each note means that each pipe can be treated by the organ builder as a separate instrument and no compromises are necessary from the acoustical point of view. However, organs are so incredibly expensive to build that compromises are often introduced purely in order to save money. For example what appear to be two different ranks of pipes controlled from different keyboards, may in fact be one rank controlled in two different ways. Also because reed pipes are far more complicated than flue pipes the upper one or two octaves of a reed rank may actually consist of flue pipes. This is possible because the essential difference in tone quality is in the number of higher harmonics present. For the higher pitched pipes, the upper harmonics are so high in pitch that they come close to the upper limit of audibility and hence their presence or absence is hardly noticed. At the lower pitches there is a considerable difference in the starting transient for flue and for reed pipes but this also tends to disappear at the higher frequencies.

Before we leave the subject of organ pipes there are three special categories that ought to be mentioned in which, for each note, several pipes are involved. The first is the category typified by the Voix Céleste. This is a rank of pipes which are tuned slightly flat or sharp relative to another rank, usually of similar quality but louder. The difference in frequency is usually about 7 or 8 Hz, so that when the two pipes are sounded together the amplitude varies at the difference frequency to give an effect similar to vibrato (see section 1.16 on beats and difference tones).

The second category is that of the so-called 'resultant bass' pipes. The longest pipes on a large organ may be very difficult to fit into a building. To produce the notes of pitch corresponding to these very low notes an alternative is to use two pipes placed back to back of considerably higher pitch than the required note but with a pitch difference that will give a resultant difference tone at the required pitch (see again section 1.16).

The third category is that of mixtures for which there may be anything from two to seven pipes assigned to each note. These pipes correspond to a group of high harmonics of the actual note required. For example the simplest mixture might be of two pipes corresponding to the third and fifth harmonics. Mixtures are never used alone but can add greatly to the richness of the sound produced by other pipes of normal pitch. The number of pipes involved for each note is specified in an organ specification by Roman numerals. For instance 'Mixture IV' means that four ranks of pipes are involved.

4.21 THE MECHANISM OF AN ORGAN

There are almost as many different ways of putting organ pipes together and of organising their control as there are organs. The organ is perhaps the only musical instrument that is usually built to fit the building in which it is to be played. There are, of course, small portable organs, or chamber organs that exist as an entity that can be carried around from place to place, but the kind of organs that I propose to describe here are very much fixtures in a particular place. Nevertheless whatever the size the basic principle of

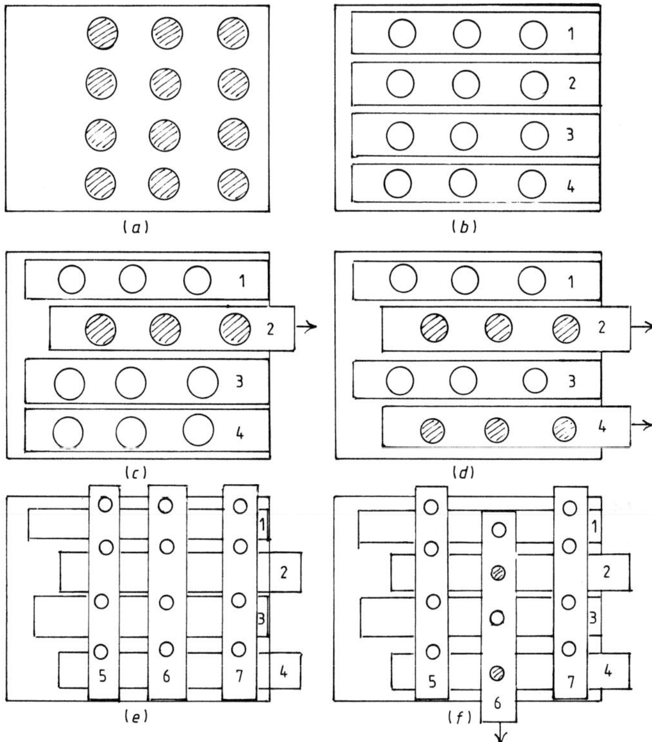

Figure 4.29 Schematic block diagram of the control mechanism of an organ showing: (*a*) the top of the wind chest with a few representative air holes; (*b*) the sliders corresponding to four different stops; (*c*) one stop pulled out so that air can emerge through the holes in slider No 2; (*d*) a second stop pulled out; (*e*) the addition of sliders corresponding to some of the keys; (*f*) a key depressed allowing air to emerge from two holes; the pitch of the note is defined by key-slider 6 and the quality by stop-sliders 2 and 4.

operation is similar. The organ of the Church of St. James, Spanish Place, in London which was used as an example in the televised lectures is really four separate organs. Three have manual keyboards that are arranged one above the other and the fourth is controlled by the pedal keyboard. At the beginning of section 4.19 we discussed the three main groupings of organ pipes and it is interesting to note that in many of the simpler organs the actual control mechanism closely resembles the schematic diagram of figure 4.29. In any one of the four organs depression of a particular key admits air to all the pipes in that organ corresponding to that note. Then, however, air will only reach the pipes if, in addition, the stop controlling that particular rank of pipes is drawn. The actual note heard will depend on the pipes drawn. For example if the key C_4 (middle C) on the middle keyboard (which controls the great organ) is depressed and stops labelled Diapason 8', Principal 4', Twelfth 2 2/3', and Fifteenth 2' on the great organ are drawn, then the actual notes sounding will be C_4, C_5, G_5 and C_6. This organ is called the 'great' because it contains the basic stops providing the body of sound most characteristic of an organ. The volume is controlled by adding or taking away ranks of pipes by using the stops. Because this sometimes has to be done when the organists hands and feet are relatively fully occupied in playing, a series of 'pistons' is provided. These are buttons under the keyboards that can be pressed with a finger or thumb, or there are also larger buttons that can be pressed with the feet. When a particular piston is pressed a group of stops is automatically drawn. Sometimes the grouping is fixed and as successive buttons are pressed more and more pipes are brought in and hence the volume can be built up or conversely brought down again. In the organ being described the pistons can be preset by the organist before the recital so that given combinations of stops can be brought in as required for the piece being performed.

The upper keyboard controls the 'swell' organ. In this case the whole organ is enclosed in a box on the front of which are heavy shutters rather like Venetian blinds which can be opened and shut by a pedal to give a finer control of the volume than is possible by adding and taking away stops. The pipes on this organ are often of the kind suitable for playing solos that can be accompanied on one of the other organs.

The third, and lowest, keyboard controls the 'choir' organ which is also enclosed in a box with shutters. It contains pipes that are generally quieter than those of the great and are used for accompanying choir singing.

The remaining controls that occur are the 'couplers'. These link together notes on the various organs. For example 'swell to great' means that any note played on the great organ will also play the same note on the swell organ; 'great to pedals' means that when a pedal key is depressed the note on the great will be sounded; and 'great suboctave' means that when a key on the great is depressed, the note an octave below will also be heard.

There are three main ways of achieving the actual control. The earliest is purely mechanical and is known as 'tracker'. In this case there are direct mechanical connections between all the various keys, stops, couplers, etc. As a result when the full organ is in use considerable effort is required to depress a key. On the other hand some organists like this because the control is so direct. Historically the second method of operation is pneumatic. When a key is depressed air is admitted to a pipe at the other end of which a little bellows is inflated and this operated the valve admitting air to the pipe. Such systems involve enormous lengths of pipe, but they provide a much lighter action and the console can be much further away from the organ pipes. The third method is electrical. It may be electromechanical, involving relays and switches, or in more recent organs various systems involving electronic devices have been used. The enormous organ in the Sydney Opera House has a computer assisted control mechanism which allows the organist to play a piece, sit down in the auditorium, set the organ to play over the piece automatically and to change the registration (i.e., the particular combination of stops in use) with a remote control as the piece progresses. This of course surmounts one of the biggest difficulties faced by organists—that they are rarely in a position to hear the composition as the audience will hear it.

4.22 THE VOICE

The human voice can be said to belong to the second family of musical instruments (section 2.6) since all the pitches required are produced by altering the parameters of the same system. The primary source of sound—whether we are speaking or singing—is the vocal chords. These are just a pair or membranes with a gap between, which create a relatively raucous sound when air is forced between them in much the same way that children produce a 'squawk' by blowing two blades of grass stretched between the thumbs. Another model for the vocal chords is an inflated balloon that is allowed to deflate with the neck stretched to make a narrow gap. The voice obviously belongs to the reed family but the shape of the cavities that are involved—the throat, mouth, sinuses, etc, are very much more complicated and capable of far more variations in shape than any conventional instrument.

You can hear the basic sound of the vocal chords if you say 'Ah'—as when requested by a doctor and the pitch of the sound can be raised or lowered by a subconscious tightening or slackening of the chords. This change of pitch by changing the tension of the chords is used to introduce liveliness into speech. Try saying out loud 'Are you in a car' in a surprised tone of voice and compare the pitches of the two 'Ah' sounds.

Many people seem to be surprised to realise that even plain speech (as opposed to singing) makes use of variations in musical pitch. In fact we shall see a little later in this section that we need to have the ability to make sounds of certain specific pitches when speaking normally. This always comes as a surprise to those who claim to be 'tone deaf'.

> A simple demonstration of the close relationship between speech and pitch can be done by pretending that you have lost your voice completely and can only whisper. This means in fact that you have stopped the vibrations of the vocal chords by letting them become completely relaxed. Then try to whisper a long 'Ah' sound and then, without stopping the flow of air, change to 'Ooh' (as in 'soup'). If you do this several times you should be able to sense a change of pitch in the breathy sound you are

producing. If you have a musical ear you may be able to judge that the change in pitch is somewhere between a fifth and an octave, depending on the exact choice of vowel sounds that you make.

In the earlier demonstration we changed the pitch of the 'Ah' sound by altering the tension of the vocal chords; now we are changing the pitch by altering the shape of the cavities of the nose, throat and mouth. Both pitch changing techniques are used in speech and singing.

The basic wave emerging from the vocal chords has a very 'spiky' form, almost like a sequence of pulses and is therefore capable of exciting a wide range of frequency responses (see section 2.13). There is also a good deal of white noise from air turbulence. This compound wave then passes through a series of tubes (the larynx, etc) to which a series of cavities (the sinuses, mouth, nostrils, etc) are all attached and is finally radiated from an aperture of variable size and shape (the mouth and lips). All kinds of changes in the shape of these cavities are possible, by muscle control, by changing the position of the tongue, by changing the shape and position of the soft palate, or by changing the position and shape of the lip aperture. There are also less controllable changes due to infection or by the accumulation of mucus. The overall change in the frequency pattern of the resultant sound can be described as the imposition of formant characteristics (see section 2.7). There are four distinct types of formant that need to be considered. First there is a more-or-less fixed one that is characteristic of the person speaking and helps us to recognise the voice. This may, of course, change with illness, particularly if it involves accumulation of mucus in some of the cavities. Secondly there is a controllable formant that deliberately changes the quality of the voice from the harsh rasp of anger, to the gentle murmur of a mother soothing a child, or to the dulcet tones of lovers. Thirdly there is the controllable formant that determines the nature of the vowel sounds. And finally there is a formant which, it is claimed, changes the speaking voice into the singing voice.

There are many interesting experiments that can be done to demonstrate the existence of voice formants and to show how they may be controlled. In section 3.3 we described an experiment in which an electric razor is used to excite resonance in the mouth of the demonstrator. The vibrating head of the razor creates a pulse-like waveform that is rich in harmonics. It is placed on the outside of the cheek or throat and the resultant sound coming from the open mouth is picked up by means of a microphone, amplifier and loudspeaker so that it can be heard by the audience. The cavity of the mouth behaves like a Helmholtz resonator and the harmonics of the razor vibration frequency can be amplified one by one. The easiest way to change the shape of the mouth cavity is to say one of the vowels silently and to exaggerate the necessary shape of the lips. The primitive instrument known

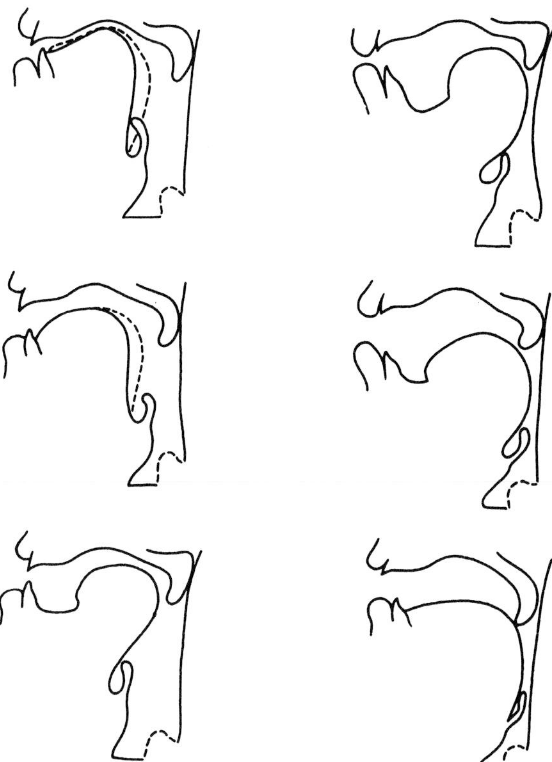

Figure 4.30 Tracings taken from x-ray pictures of the vocal tract while articulating a series of six Russian vowels.

as the Jew's harp, or Guimbard has a metal tongue which, when plucked produces a basic sound and various components of it can be amplified by the mouth cavities in the same way in order to play tunes (section 3.3).

The well known change in voice quality that arises when helium gas is breathed, as for example in certain diving operations, arises because when these cavities are filled with helium, for which the velocity of sound is almost three times that of air, the resonant frequencies are changed. The vocal chord frequencies are unaffected and the combination is quite unnatural.

Some of the earliest work on vowels was done by Sir Richard Paget in the 1930s. He used a 'squeaker' driven by air from a bellows to represent the vocal chords and added cavities made of cardboard and plasticene to model the cavities of the nose and throat. A set of his models still exists in the Royal Institution in London and is still occasionally used to demonstrate vowel formants. Among the many more recent studies is the highly sophisticated work of Professor Gunnar Fant of the Royal Institute of Technology in Stockholm. His studies, based on x-rays of the vocal tracts of people while producing sounds are classics, not only in relation to the results, but also for the elegance of the methods used. Figure 4.30 shows tracings of vocal tracts made from Professor Fant's x-rays of people speaking Russian vowels. The point that is so surprising about vowel formants is that the frequencies associated with the different vowel sounds are absolute. There is always some debate on exactly how many formant regions there are but it is quite certain that three predominate. One or two higher ones have been claimed by some workers.

In order to bring out some of the important features of the mechanism of vowel production a series of experiments were set up by the author for the 1971 Christmas Lectures and, since the results still make useful points they are repeated here. All vowel sounds begin with the buzz of the vocal chords; this is a harmonic-rich sound with a fundamental which in men lies between 50 and 250 Hz and in women and children can be at least an octave higher. The harmonics are amplified by the formant characteristics and in the experiment to be described only the three principal formant regions were invoked. The interesting point is that, although the vocal chord frequencies vary, depending on, among other things, the age and sex of the speaker, the formant regions for a given vowel do not.

Six speakers were invited to pronounce a set of six vowels. The speakers were a man from the North of England, a man from Wales, a woman from Wales, a woman from Czechoslovakia, and a girl and a boy both from Wales. The vowels used were Ah, Aw, Er, Eeh, Oh and OOh. The sounds were analysed using a B & K real-time frequency analyser (see section 2.12) and the results plotted as graphs which are shown in figure 4.31. Many points can be illustrated by these figures but perhaps the two most important are first, that there is a strong similarity between the graphs for all six speakers of a given vowel in the middle region of the graphs (500–200 Hz); this, of course, demonstrates the point that the vowel formants are independent of the speaker. Secondly, if you compare the graphs for a given speaker there is a strong similarity, especially at the lower end; this demonstrates the point that there are formants characteristic of the person that are independent of the vowel.

In more recent years the suggestion has been made (especially by workers in Sweden) that an additional formant in the region of 3500 Hz is involved when speech becomes singing. It has also been suggested that this may be cultivated by certain techniques in voice training because long time average spectra (that is the power at each frequency averaged over a period of a few minutes) for a whole orchestra tend to have a lower power level at this kind of frequency than at higher or lower frequencies. Thus an opera singer should be able to be heard more easily above the orchestra if such a formant exists.

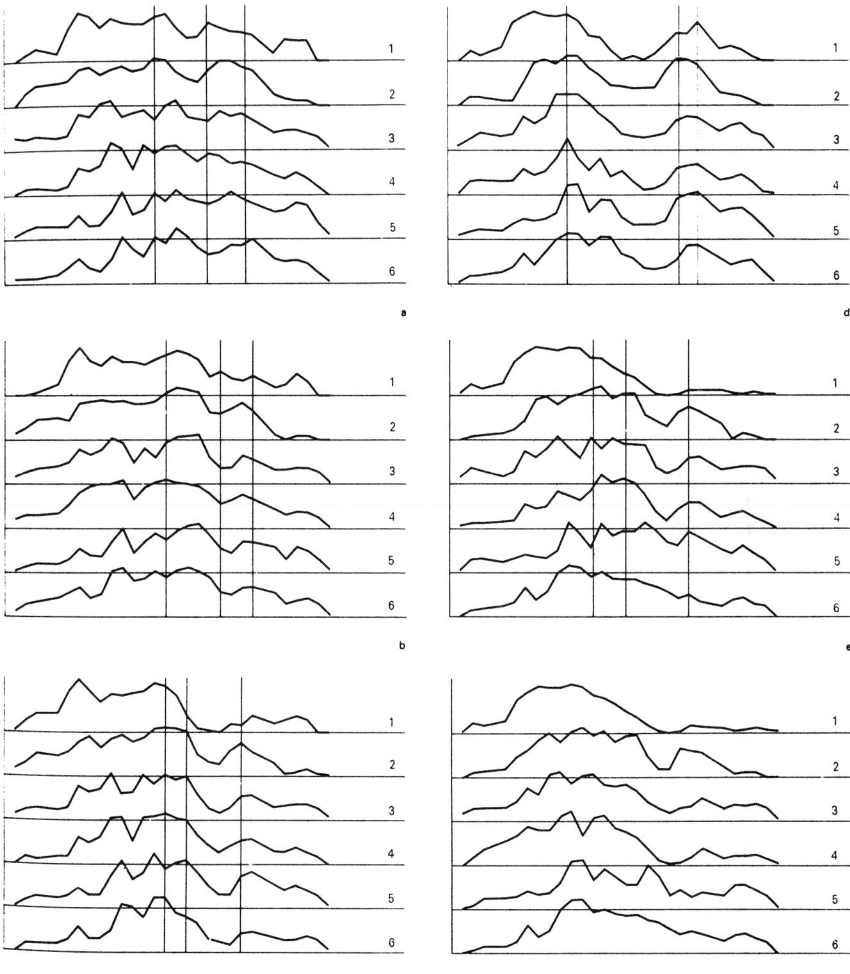

Figure 4.31 The frequency distribution of six different vowels spoken by six different people. The vowels are: (*a*), Ah; (*b*), Aw; (*c*), Er; (*d*), Ee; (*e*), Oh; and (*f*), Ooh. The three vertical lines are the generally accepted main formant frequencies. The people are: (1) man from North of England; (2) man from Wales; (3) woman from Wales; (4) woman from Czechoslovakia; (5) boy from Wales; (6) girl from Wales.

The whole question of voice training is a fascinating one that is well beyond the scope of this book. However, it may be worth mentioning in passing that the basic problem is to enable singers to place given muscles in the right states of tension even though *conscious* control of each individual muscle is not normally possible. The various verbal descriptions that may seem contradictory such as 'sing from the head', 'place the voice in the chest', etc, are all complementary methods of arranging for the muscles to do what is required.

4.23 CONCLUSION

In this chapter we have followed a similar course to that of Chapter 3 but with wind instruments instead of strings. In many ways the divergence of the behaviour of real instruments from that of tubes in the laboratory is even greater than that for strings. We have seen particularly how important an influence on music the actual technology of instrument making has had. The ingenious Boehm system of keys for the woodwinds made possible the playing of complicated passages that would be almost impossible to play on early instruments. The invention of valves and slides made it possible for brass instruments to play tunes in their lower registers and so influenced the compositions that were written for them.

Two of the greatest of all wind instruments—the voice and the organ—have been included in this chapter, but the exotic electronic organs have been moved into the next chapter with other electronic instruments.

5

Scales, Synthesisers and Samplers

5.1 INTRODUCTION

In this chapter we shall be concerned almost entirely with instruments belonging to the first family; that is instruments that have a separate source for every note. In most cases they are controlled by a keyboard. But keyboard instruments (including the harpsichords, pianos and organs discussed at the end of Chapters 3 and 4) depend on the possibility that a series of useful notes can be arranged in some sort of methodical way so that they can be found easily by the player and produced at will in order to play tunes. The scheme that is used is to place the notes in some sort of scale, or series of steps. But the big question is how to choose the particular scale to be used. Many different scales have been proposed and many different ones still exist in different parts of the world. The reason for including the subject of scales in this chapter concerned principally with electronics, is that it is only since the advent of electronic instruments that it has become possible to reproduce many different kinds of scales on one instrument and to switch rapidly from one to the other in order to hear the differences.

Some of the differences are very small indeed and, while I understand the theoretical reasons for the choice of some of them, I find it difficult to believe that anyone but the most highly trained musicians could really distinguish between some of them. I suspect too that all tuners would not be capable of setting particular instruments in a particular scale without the aid of electronics. How it was done in the time of Helmholtz I find it difficult to imagine.

The first point to get clear is that music comes first and scales later. Scales can be compared with the grammar of a language; it is perfectly possible to speak a language for the whole of one's life without ever understanding its grammar. And, in the same way, musicians, especially those in the folk tradition, can write, play and sing musical compositions without being consciously aware of the scale structure involved.

This is neither a textbook on musical theory, nor an account of the history of music, and so I propose to treat scales as they are now and to concentrate on the complications that they introduce into the design of keyboard instruments.

5.2 THE PURPOSE OF SCALES

If the purpose of an instrument is to play tunes using single notes then the scale is not as important as if it is to be used to play chords. Since one of the main advantages of keyboards is the possibility of a single person playing chords it follows that scales are of the greatest importance to keyboard players.

The basic problem is to define a set of frequencies that will blend together in a harmonious way. In section 1.15 we saw that this means that the frequency ratios must be simple. Obviously there must not be too many separate notes or our keyboard will become unmanageable, and the intervals between the closest notes must be sufficiently large so that even an untrained ear will be able to distinguish them.

One useful way of deriving such a series is to use the two simplest ratios, that is 2:1 (the octave) and 3:2 (the fifth). Suppose we take a base note which we will define as having a frequency of one unit and then consider notes that are a fifth above and a fifth below. This give us three different notes and we will then take the first 5 harmonics of these three notes. These notes will be very widely dispersed, but we can bring them within the span of one octave by using the octave ratio, 2:1 or 1:2; That is we divide or multiply successively by 2 until the note lies within the chosen octave.

Table 5.1 shows how this works out in practice.

Table 5.1

Harmonics on the note a fifth below the base note, the base note and a fifth above the base note:-

$$\tfrac{2}{3} \quad \tfrac{4}{3} \quad \tfrac{6}{3} \quad \tfrac{8}{3} \quad \tfrac{10}{3} \qquad 1 \quad 2 \quad 3 \quad 4 \quad 5 \qquad \tfrac{3}{2} \quad \tfrac{6}{2} \quad \tfrac{9}{2} \quad \tfrac{12}{2} \quad \tfrac{15}{2}$$

Place in ascending order and eliminate duplicates:-

$$\tfrac{2}{3} \quad 1 \quad \tfrac{4}{3} \quad \tfrac{3}{2} \quad 2 \quad \tfrac{9}{2} \quad \tfrac{8}{3} \quad 3 \quad \tfrac{10}{3} \quad 4 \quad 5 \quad 6 \quad \tfrac{15}{2}$$

Divide or multiply successively by 2 to bring within the octave on 1:-

$$\tfrac{4}{3} \quad 1 \quad \tfrac{4}{3} \quad \tfrac{3}{2} \quad 2 \quad \tfrac{9}{8} \quad \tfrac{4}{3} \quad \tfrac{3}{2} \quad \tfrac{5}{3} \quad 1 \quad \tfrac{5}{4} \quad \tfrac{3}{2} \quad \tfrac{15}{8}$$

We now collect up all the new ratios that have been produced and arrange them in order and we have:

Table 5.2

1	9/8	5/4	4/3	3/2	5/3	15/8	2

This sequence obviously obeys the criteria that we set out at the beginning of the section—seven steps is a reasonable number and the intervals are easy to distinguish even by an untrained ear. If we multiply them all by 264 Hz we find in fact that they give us an approximation of the scale of C, that is a scale starting on C and using only the white notes on the piano. (You will see why I say approximately a little later when we discuss equal temperament.) The reason why we chose the frequency of C_4 to be 264 Hz is that it leads to A_4 being 440 Hz which is the standard pitch used by orchestras.

Table 5.3

264	297	330	352	396	440	495	528
C_4	D_4	E_4	F_4	G_4	A_4	B_4	C_5

This is the scale that is usually called the 'diatonic major' scale and is the familiar sequence 'doh, ray, me, fah, soh, lah, tee, doh'. It came into being during the 17th century together with the diatonic minor scale (which we shall discuss briefly in the next section). Before that time music had been related to the eight Gregorian modes, each of which has a different sequence of intervals, and which are still sometimes used, especially in Church music.

Now that is fine if we only want to play in the key of C (i.e., using the scale beginning on note C). But suppose we want to play in a scale beginning on D. That is, we now call the note D 'doh' and move up with the same intervals as before. To do this we can simply multiply all the frequencies of the scale of C (table 5.3) by 9/8 (which is the ratio of the frequencies of D/C). The result is then:

Table 5.4

297	$\underline{334.1}$	$\underline{371.2}$	396	$\underline{445.5}$	495	$\underline{557}$	594
D_4	(E_4)		G_4	(A_4)	B_4		D_5

Two of the underlined notes (334.1 and 445.5) are close to, but not exactly the same as notes E_4 and A_4 in the scale of C (table 5.3) above; they are out of tune and we have indicated this by the brackets:

The other two underlined notes are about midway between notes from the scale of C (F and G, C and D) but slightly nearer to the lower of the two; we will call these F sharp and C sharp (written $F^{\#}$ and $C^{\#}$).

Table 5.5

297	$\underline{334.1}$	371.2	396	$\underline{445.5}$	495	557	594
D_4	(E_4)	$F^{\#}_4$	G_4	(A_4)	B_4	$C^{\#}_5$	D_5

Now let us repeat the procedure starting on E; that is we multiply the basic scale by 5/4 and we arrive at:

Table 5.6

330	371.2	<u>412.5</u>	440	495	<u>550</u>	<u>618.8</u>	660
E_4	$F_4^{\#}$	$G_4^{\#}$	A_4	B_4	$(C_5^{\#})$	$D_5^{\#}$	E_5

Again we have introduced new notes (underlined) two of which are midway between existing notes and can be dealt with by using sharps; but the third (which we have labelled $C_5^{\#}$) is close to, but not exactly the same frequency as that of $C_5^{\#}$ in the scale of D.

We will perform the same operation just once more, starting with the note F, by multiplying the frequencies of the scale on C by the ratio F/C, i.e., 4/3.

Table 5.7

352	396	440	<u>469.3</u>	528	<u>586.7</u>	660	704
F_4	G_4	A_4	$B_4^{\flat}$	C_5	(D_5)	E_5	F_5

Again there are two new notes; one of them (586.7) is almost, but not quite D_5, so we have labelled it so with a bracket to indicate that it is out of tune. But 469.3 is midway between A_4 and B_4 (440 and 495 Hz), but this time it is closer to the higher of the two notes, i.e., B_4, so we will designate it B flat (written $B^{\flat}$).

The significant point is that while we can handle the notes that are roughly midway between the notes of the diatonic scale we run into problems with the 'near misses', or out-of-tune notes. If they are used in chords there would obviously be some violent discords.

This scale, which works perfectly in any one given key is called 'Just' intonation (labelled J in later tables).

5.3 EQUAL TEMPERED SCALES

Let us now examine the actual intervals between the notes of the diatonic scale.

$$D/C = \tfrac{9}{8} \qquad E/D = \tfrac{5}{4}/\tfrac{9}{8} = \tfrac{10}{9} \qquad F/E = \tfrac{4}{3}/\tfrac{5}{4} = \tfrac{16}{15}$$
$$G/F = \tfrac{3}{2}/\tfrac{4}{3} = \tfrac{9}{8} \qquad A/G = \tfrac{5}{3}/\tfrac{3}{2} = \tfrac{10}{9}$$
$$B/A = \tfrac{15}{8}/\tfrac{5}{3} = \tfrac{9}{8} \qquad C/B = \tfrac{2}{1}/\tfrac{15}{8} = \tfrac{16}{15}$$

So we have only three different intervals:

Table 5.8

C	9/8	D	10/9	E	16/15	F	9/8	G	10/9	A	9/8	B	16/15	C

and it is this sequence of intervals that characterises the diatonic scale starting on whatever note.

Many ingenious systems have been devised to get round the difficulty because sticking to these intervals in all keys produces far more notes than can normally be incorporated on a keyboard.

The most useful and most universally used is that known as 'equal temperament' (labelled ET in later lists). The intervals between the notes are 'tempered' to fit and although all the notes as a result of the compromise are a little out of tune the result is acceptable in all keys. It is often said that J S Bach wrote his famous 48 preludes and fugues (two in each of the 24 possible keys) to illustrate how well the system works. In fact, more recent studies do not seem to support this view and it seems more likely that Bach used a temperament of his own. Hence the name of the set is the 'well-tempered keyboard'.

Instead of sticking to the three different kinds of intervals (9/8, 10/9 and 16/15) we make the approximation that 9/8 and 10/9 are in fact the same interval, called a tone; and we call 16/15 a semitone and make two semitones equal to a tone.

Then the interval sequence in the diatonic scale becomes:

C tone D tone E semitone F tone G tone A tone B semitone C

and if we try this on the scales of D, E and F above it still works.

D	tone	E	tone	F#	semitone	G	tone	A	tone	B	tone	C#	semitone	D
E	tone	F#	tone	G#	semitone	A	tone	B	tone	C#	tone	D#	semitone	E
F	tone	G	tone	A	semitone	B$^\flat$	tone	C	tone	D	tone	E	semitone	F

Thus we need to divide each octave on the keyboard into twelve equal semitones. The interval corresponding to each semitone must then be the twelfth root of 2, that is 1.059463.

So the equal tempered diatonic scale starting on C_4 (middle line) compared with the just scale (bottom line) becomes

Table 5.9

C_4	D_4	E_4	F_4	G_4	A_4	B_4	C_5	
261.6	293.7	329.6	349.3	392	440	493.9	523.2	ET
264	297	330	352	396	440	495	528	J

that on D_4 becomes

Table 5.10

D_4	E_4	$F_4^{\#}$	G_4	A_4	B_4	$C_4^{\#}$	D_5	
293.7	329.6	370	392	440	493.9	554.4	587.4	ET
297	334.1	371.2	396	445.5	495	557	594	J

that on E_4 becomes

Table 5.11

E_4	$F_4^{\#}$	$G_4^{\#}$	A_4	B_4	$C_5^{\#}$	$D_5^{\#}$	E_5	
329.6	370	415.3	440	493.9	554.4	622.3	659.2	ET
330	371.2	412.5	440	495	550	618.8	660	J

and that on F_4 becomes

Table 5.12

F_4	G_4	A_4	$B_4^{\flat}$	C_5	D_5	E_5	F_5	
349.3	392	440	466.2	523.2	587.4	659.2	698.6	ET
352	396	440	469.3	528	586.7	660	704	J

You will notice:

(i) that in the equal tempered scale there is now no distinction between sharps and flats as both are exactly midway between adjacent white notes; this is of course a necessity if the conventional piano keyboard is to be used;

(ii) that the discrepancies between the equal tempered frequencies (ET) and the just frequencies (J) are all very small.

The other scale that is particularly important in Western music is the diatonic minor scale which is more closely related to some of the modes of early Church music than is the major scale. The difference is in the sequence of intervals. We shall concern ourselves only with the equal-tempered version.

The interval sequence within the octave in major scales is the same whether the scale is rising or falling:

tone, tone, semitone, tone, tone, tone, semitone

However, for the minor scale a distinction is made between a rising scale and a descending scale. The sequences are:

rising: tone, semitone, tone, tone, tone, tone, semitone
descending: tone, tone, semitone, tone, tone, semitone, tone.

The aesthetic reason for the difference that is usually given is that, in descending, if the same scale as for rising were used, it would sound exactly the same as the major scale until the sixth note was reached. However, my own view is that, for the purposes of this book it would be unwise to delve further into these esoteric mysteries and simply accept the scales as facts!

I mentioned earlier that Bach wrote the '48' to demonstrate that a good temperament will work in all the 24 possible keys. The number 24 is arrived at by assuming that there is one key starting on each of the 12 notes of the octave and that each can be either major or minor.

5.4 CONSEQUENCES OF TEMPERAMENT

Many volumes have been written on temperaments and on ways of tuning keyboard instruments to fit various schemes. The problem becomes most acute in the tuning of organs because of the sustained nature of the tones. From the middle of the 19th century for about a hundred years it was fashionable to tune organs to equal temperament. But the second half of the 20th century has seen a gradual movement away from this. Charles Padgham in 1986 listed about a hundred organs in England that are no longer tuned in equal temperament. It would be out of place in this book to delve more deeply into the rights and wrongs of the many other temperaments that are possible and which for some purposes work better than the middle of the road compromise of equal temperament. Charles Padgham points out that being able to move from one key to another without disaster is not a feature limited to equal temperament. Indeed he points out that, as mentioned in the last section, Bach's '48' were described as for the 'well-tempered keyboard' and suggests that if Bach had intended them to be played with equal temperament he would have said so! I shall leave those arguments for those far better qualified than me to pursue.

However, there are several consequences of different schemes of temperament that we need to consider and the first of these is whether players of non-keyboard instruments use just intonation or whether they play in equal temperament. I am prepared to make the somewhat heretical statement that they vary from time to time depending on the nature of the piece being performed and on the nature of the other instruments or voices that are also involved. My guess is that they instinctively play notes that sound 'right' in the context and may be at least partially unaware of the particular scales being used. This fits in well with the statement I made at the beginning of the chapter to the effect that scales are formalisations of structure that are produced after the music has been established.

But, of course, with keyboard instruments the player has no control; the temperament is established by the tuner. And it is not too surprising to find that there are many 'rules of thumb' methods of 'laying out the temperament' which can be used without the tuner ever being aware of precise frequencies or indeed of any physical measurements. Among other techniques, listening for beats between various harmonics, counting beats to establish divergence from just intervals and so on can be used. The ultimate test is not 'Do the frequencies fit the theoretical requirements of the chosen temperament?' but rather 'Does the finished instrument sound good to the player and to the listeners?' The obvious corollary is that the primary requirement of a good tuner is to have a good 'musical ear'.

The advent of electronic keyboard instruments has made it possible to experience the effects of different temperaments very quickly. Using the Yamaha DX7IID synthesizer in the lectures I was able to demonstrate some of the more obvious consequences. For example we first of all set the synthesizer to play in just intonation on the key of C as in table 5.3 and play various pairs of notes and chords using fifths, fourths and thirds. The sounds are all pleasant. Now, without changing the temperament we play a similar set of chords in the key of D. The discordant nature is instantly apparent to the ear.

Now the same demonstration is repeated with the keyboard set to equal temperament. Now the chords in the two keys are both somewhat discordant but neither is as bad as those in the key of D in just intonation.

Now we have introduced a very modern synthesizer but before dealing with it in more detail we need to consider how electronic synthesizers developed.

5.5 ELECTRONIC SYNTHESIS

It is possible to identify four separate strands in the development of electronic synthesizers, although they tend to overlap. They do, however, provide a structure on which to build the next sections of this chapter.

The first strand is that of analogue synthesis and it began quite early in the present century. Indeed the forerunner in terms of theory, although it never became a practical project, was Cahill's 'Telharmonium'. The idea was to use, in effect, a series of rotating dynamos, each of which would generate alternating electric currents at different frequencies. In terms of the three families of instruments that we have used earlier this would belong to family number one since it had a separate generator for each note. These would then be selected and mixed in various ways by a performer and the resulting current delivered into the telephone network so that subscribers could hear the music. The basic principle worked but there was so much interference with the rest of the telephone service that it could not be used. In some ways it could be thought of as the precursor of the Hammond and Compton electronic organs that appeared in 1932. Other instruments in the analogue category were devices like the 'Theremin' and the 'Ondes Martenot' which belong to family number two in that they used a single oscillator whose frequency could be varied to permit tunes to be played using the resultant electrical oscillations to drive a loudspeaker. Since the 1970s many other analogue synthesizers have appeared that really belong to both families one and two in that the basic tone generation is by oscillators whose frequency can be varied, but then they can be grouped together, mixed and modified from one or more keyboards like an organ.

The second strand begins with the technique known as '*musique concrète*'. It depends on the tape recorder. Real sounds are used: they could be inherently musical, like a voice or the sound produced by blowing across the neck of an empty a bottle, or they might be in the category of noise like the barking of a dog or the clang of a dropped spoon. Once recorded the sound can be played back at different speeds to give a whole series of musical notes and then the composition assembled by literally cutting up the tape and splicing it together again. Although incredibly time consuming it enjoyed considerable vogue especially for background music in films and television. It has now almost disappeared and has been replaced by another system which is based on a sophisticated electronic method of performing the same functions with enormous increases in speed and flexibility. It still starts with real sounds but they are recorded in digital form and then can have their frequencies and other aspects changed and can be assembled entirely electronically.

The third strand is that of digital synthesis. It can be traced back to the very earliest digital computers that were made to perform party tricks for the mystification of visitors. For example the frequency at which the output in the form of punched tape was produced could be varied in order to play a simple tune. Nowadays, however, the ramifications of digital synthesis using either a full computer system with specially designed software, or specially designed keyboards with in-built computers are endless.

The fourth strand begins with the concept of voltage control in which, for example, the frequency of an oscillator in an analogue synthesizer could be controlled by applying different voltages to it. This meant that the control of one element of a synthesizer by another became possible. For example the frequency of one oscillator producing a basic tone could be varied by applying voltages derived from a keyboard in order to play tunes, but also by the output of a second low frequency oscillator to produce a vibrato effect. The potential for musical composition was enormously increased by the possibilities of this kind of interconnection. It was also possible to store and play back a piece of music by storing a sequence of voltages which could later be applied to the synthesizer to reproduce the piece. The storage device is called a 'sequencer'. Voltage control is, in fact, described in the next section under analogue synthesis because this is where it was first used. The current descendant (in terms of what it can do, rather than of how it does it) is the system known as MIDI (Musical Instrument Digital Interface) which literally permits any electronic instrument to control or to be controlled by any other. The ultimate example of this in the lectures was a visit to the BBC Radiophonic Workshop to see their latest studio in which a vast array of electronic devices is controlled by a computer through the medium of MIDI.

5.6 ANALOGUE SYNTHESIS

Many of the early analogue instruments mentioned at the beginning of section 5.5 were almost in the category of instruments like the musical saw; that is they were played by specialists more as 'acts' than as musical performances. It is probably not worth spending more time on them than to mention that they consisted mainly of an oscillator whose output could be applied to a loudspeaker, together with various ingenious ways of altering the pitch and volume in order to play tunes. The first analogue synthesizers to be manufactured commercially and to become generally available (although they were not at the time described as synthesizers) were the Compton and the Hammond electronic organs that both appeared in 1932.

The Compton Electrone system uses rotary tone generators which produce tiny changes in electrical capacity that can be converted into voltage changes which, in turn, can be amplified for use. In the simplest organ of this type there would be twelve tone generators, one for each note of the chromatic scale, C, C$^{\#}$, D, D$^{\#}$,..., etc. Each of the tone generators produces seven notes at octave intervals (see figure 5.1) by having seven tracks of varying shape on a stator and seven rings of teeth on the rotor which rotates in proximity to the stator. Each generator is driven by systems of pulleys and belts and the rotation of each is adjusted to be in the right ratio to give all the frequencies required. Twelve generators produce the 84 basic notes that can then be selected, mixed, have transients superposed on them and so on. In larger organs more complex generators produce more complex wave forms and it is clear that a very wide range of sounds is possible.

The Hammond organ that appeared at about the same time also uses rotating tone generators but they are electromagnetic. Rotating metal discs with specially shaped edges generate varying electric currents in coils wrapped round magnets fixed close to the edge of the rotating discs (see figure 5.2). The special feature of the Hammond organ was that generators were provided for each note and also for a series of harmonics of the notes. The tone could be varied by means of sliders which varied the volume of each of the contributing harmonics. Various preset registrations could be provided, or, for the organist with some knowledge of harmonic theory, the proportion of each contributing harmonic could be set.

One of the big advantages of these organs was that, because the frequencies were all derived from the rotation of a single electric motor, they could

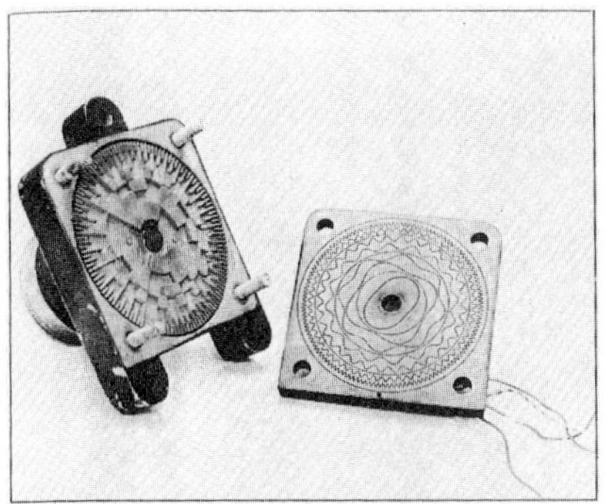

Figure 5.1 An example of the variable capacitance tone generators in a Compton Electrone organ.

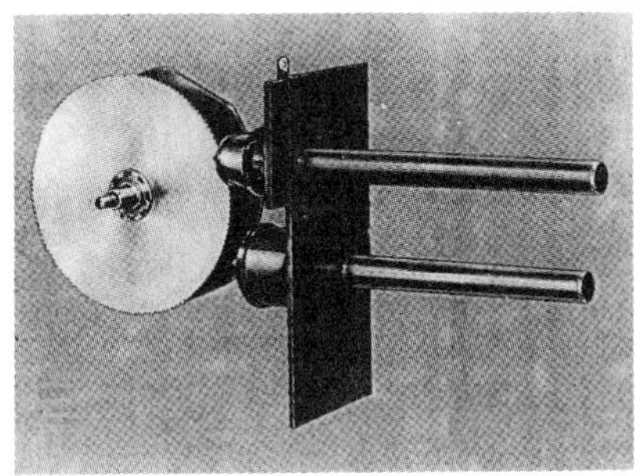

Figure 5.2 An example of the electromagnetic tone generators in a Hammond organ.

not go out of tune. Indeed by varying the speed of the motor the pitch of the whole organ could be raised or lowered by simply altering the motor speed.

Great ingenuity in circuit design very quickly gave these organs an established place as serious musical instruments. But an interesting philosophical problem soon arose, to which we shall return at the end of the chapter. It revolves round the question of whether to regard electronic instruments as means of imitating 'real' instruments, or whether they should be regarded as instruments in their own right.

After the Second World War there were many other kinds of electronic organs, many of which used electrical oscillators to produce basic tones that could be mixed in various ways. Others used as basic generators oscillators that produced very 'spiky' waveforms that were very rich in harmonics and developed the different tone qualities by filtering out various components.

As was discussed in section 2.18, almost all these early organs suffered from the same major defect. The sounds produced were too perfect, particularly in the way their amplitude varied with time, with a result that the brains of the listeners identified them as 'machines'. A steady note really was steady with no variations in amplitude whatever; any vibrato introduced was absolutely regular; and, perhaps most significant of all, the harmonic recipe for every note note of a given 'stop' was identical to that of every other note.

As mentioned earlier, none of the instruments so far described in this section was actually called a synthesizer, because the format of the instruments was very much that of a pipe organ. The earliest instruments to be called synthesizers began to appear round about the middle of the 1950s and the word is usually held to have been coined by the American company RCA. They were not single instruments in the sense that that the

Figure 5.3 General view of the lecture theatre of the Royal Institution during the 1971 series of Christmas lectures. The six VCS 3 synthesizers, the sequencer and the keyboard can all be seen behind the lecturer.

electronic organs were, but rather interconnected electronic systems that together could produce music.

There were many varieties and I shall describe only one, the VCS 3, chosen partly because it illustrates many of the features of early synthesizers and partly because it is the one that we first used in my department in Cardiff round about 20 years ago. For my first set of Christmas Lectures in 1971, through the kindness of Peter Zinoviev, I was able to borrow six of these instruments and they were linked to each other and to a sequencer to form a single large synthesizer. Figure 5.3 is a general view of the theatre during the 1971 lectures in which the six synthesizers, the sequencer and the keyboard can all be seen.

Figures 5.4 and 5.5 shows in diagrammatic form how the device is used. It consists essentially of four main kinds of unit which can all be interlinked in various ways by means of a matrix or pin board. Inserting a pin into

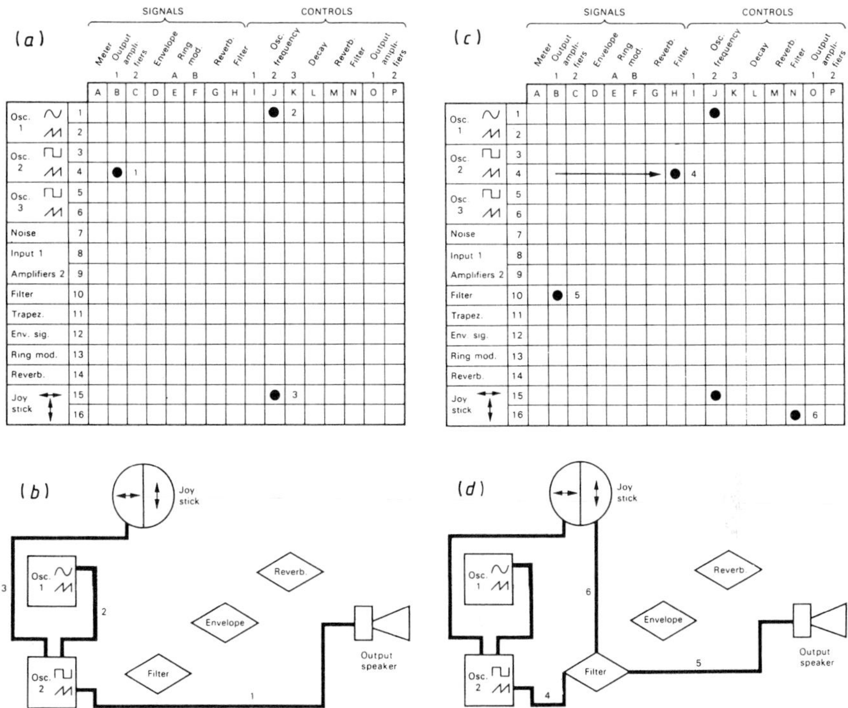

Figure 5.4 Diagram illustrating the operation of the voltage-controlled VCS 3 synthesizer. In each case the upper diagram shows how the pins are inserted into the patch board and the lower diagram shows the consequent circuit changes.

one of the holes interconnects one of the units indicated in the column at the left hand side with a musical signal or a control signal from one of the devices listed across the top.

The basic units are (1) sources of sound signals, such as oscillators, noise generators, etc; (2) treatment units, such as a filter to modify the harmonic recipe, the envelope shaper which changes the rate of rise and decay of the amplitude of the sound; (3) controllers such as the joystick, control knobs, or keyboard that provide controllable voltages which in turn can be used to control other elements in the system; and (4) interconnecting devices such as the ring modulator and indeed the pin-board itself.

The principle that makes it so powerful and yet so flexible is that of voltage control. Each of the units already listed can be controlled by applying a variable voltage; for example the frequency of the oscillator can be varied by applying a variable voltage; the rate of rise or decay of the envelope shaper can be varied by applying a voltage; and so on. The result, in addition to permitting all sorts of interconnections, is that a whole piece

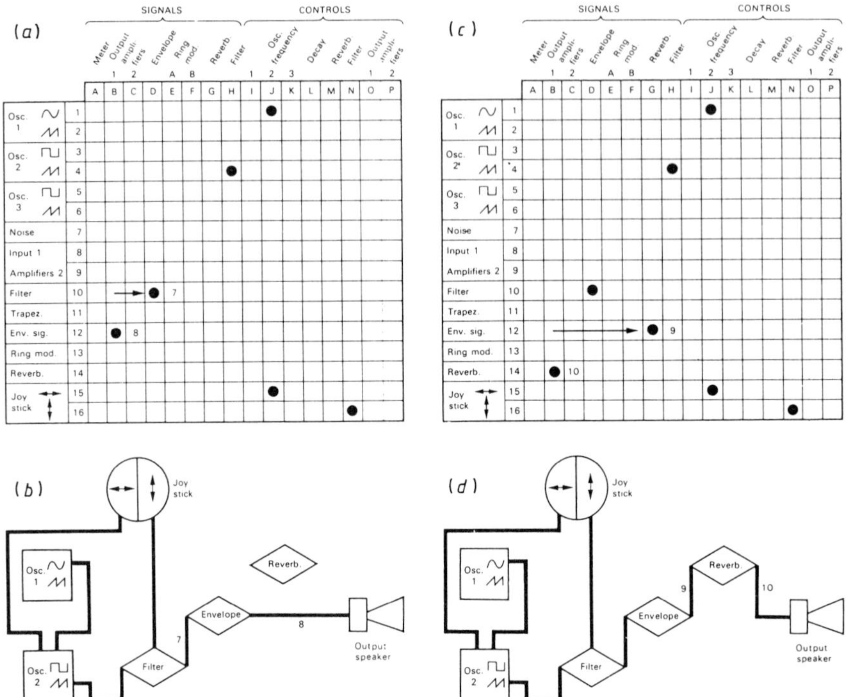

Figure 5.5 Diagram illustrating the operation of the voltage-controlled VCS 3 synthesizer. In each case the upper diagram shows how the pins are inserted into the patch board and the lower diagram shows the consequent circuit changes.

of music can be stored in the sequencer as a sequence of voltages. When these voltages are fed back into the synthesizer it is actually re-performing the piece and, in real time, it is possible to make changes in pitch, quality, etc, while the piece is in progress.

Modern synthesizers and keyboards are now so sophisticated that a high proportion of users now treat them as 'black boxes'. It has always been my contention that much more creative use can be made of black box systems if the user knows something of the processes that are going on inside. I therefore propose to describe the operation of the completely outdated VCS 3 in some detail, because no matter how sophisticated a computer programme is it can usually be thought of as performing similar basic operations using units of software in place of actual hardware as in the analogue machine. The following section in small print can be omitted by those not interested in the historical detail.

In figures 5.4 and 5.5, the conventions are that sources of signals are represented by squares; treatment units are diamonds; and control units are circular. Signals intended as sounds enter from the left and leave from the right; control signals enter at the top and leave from the bottom. We shall follow the procedure for setting up a 'patch' (that is a group of pins in the pin-board) for a simple set of operations, as it was described in the book of the 1971 lectures. Figures 5.4 and 5.5 are again involved. The upper diagram in each case shows the positions of pins as black dots; the numbers beside the dots indicate the number of the operation involving insertion of the pin and the numbers on the lower diagram indicate the circuit link that is made by the insertion of the pin.

Figures 5.4(*a*) and (*b*).

Operation 1. Insert pin at 4B. The signal from the sawtooth output of 2 is connected to the output loudspeaker and we hear a continuous, slightly reedy note, whose frequency may be controlled by turning the frequency dial on oscillator 2.

Operation 2. Insert a pin at 1J. The signal from the sine-wave output of oscillator number 1 now controls the frequency of oscillator number 2. Set the amplitude of oscillator 1, using the dial, to be small, and its frequency to about 7 Hz and the steady signal becomes modulated with a vibrato.

Operation 3. Insert pin 15J. Now left–right movements of the joystick will also control the frequency of oscillator number 2 so that its mean frequency can be set to anything we like while maintaining the vibrato.

Figures 5.4(*c*) and (*d*).

Operation 4. Remove the pin from 4B and place it in 4H. The output of oscillator number 2 now goes to the filter instead of direct to the loudspeaker.

Operation 5. Insert a pin at 10B. The output from the filter now feeds the loudspeaker and the sound we hear has been modified by the formant characteristic of the filter.

Operation 6. Insert a pin at 16N. The formant characteristic of the filter can now be changed by vertical movement of the joystick.

Figures 5.5(*a*) and (*b*).

Operation 7. Remove pin from 10B and insert it at 10D. The output of the filter now goes to the envelope shaper instead of to the loudspeaker.

Operation 8. Add a pin at 12B The output from the shaper now goes to the loudspeaker, but no sound emerges unless the 'trigger' button is pressed and then we obtain a single note whose envelope characteristic (i.e., rise, fall, duration, etc) can all be set by adjusting the control knobs on the shaper.

Figures 5.5(*c*) and (*d*).

Operation 9. Remove pin from 12B and insert it at 12G. The output from the shaper now goes to the reverberation unit instead of to the loudspeaker.

Figure 5.6 The 'Synthi 100' installed in the Physics department of University College Cardiff in early 1972.

> *Operation 10.* Insert a pin at 14B. The output of the reverberation unit now goes to the loudspeaker.
>
> In a synthesizer of this type a keyboard can be added. The keyboard produces voltage signals that can be adjusted to give notes of an equal tempered scale when these voltages are applied to the oscillator in place of the joy- stick. The depression of a key also operates the trigger in the envelope shaper.

Figure 5.6 shows a very much larger version—the Synthi 100—one of which was installed at the BBC Radiophonic Workshop and another in my department at Cardiff round about Christmas 1971.

Musical synthesis by computer was also being performed at this period but the great advantage of the analogue synthesizers was that what was stored in the sequencer was a relatively small set of control voltages for each note. In the corresponding computer operation 40,000 amplitude values were required each second in order to reproduce all the pressure variations in each note. For example, to reproduce digitally even a pure tone of constant pitch and amplitude would still require 40,000 values of the displacement every second, whereas a single parameter each for frequency and amplitude would be all that was needed in an analogue machine. At that time the saving in the necessary computer memory was very significant. Now, however, the storage possibilities even on small computers has increased so much that point by point digital synthesis is a very practical proposition.

However, there are still some special kinds of sounds that can be produced satisfactorily only by analogue synthesizers and, even in the most modern studios, analogue synthesizers are still found, but they are incorporated into the overall system by means of MIDI (see section 5.9). One such example is the sound of wind in the trees—which is very easy to imitate on most analogue computers, but much more difficult and complicated by digital methods.

5.7 SAMPLING

One of the most powerful of current techniques is that known as sampling and, in order to understand it we need to start with the much earlier system of generating electronic music known as '*musique concrète*'. The term itself was originated by Pierre Schaeffer in 1948 and was applied to 'real' sounds (as opposed, for example, to electronic sounds) that did not originate from musical instruments, but were used in musical compositions without modification. This narrow usage was soon widened to include modified sounds. One of the factors that led to the wide use of *musique concrète* in the sixties and early seventies was the tape recorder. It was actually invented before the Second World War but it did not become widely available until about 1950.

It then became possible to record real sounds (bird song, dogs barking, breaking glass, etc) and to play them back at different speeds, so changing the pitch of the resultant sounds. The tape could then be literally cut up and spliced together to give pitches in the right sequence to produce a musical composition. It is to this kind of operation that the term *musique concrète* is now usually applied.

In the late 1960s the BBC Radiophonic workshop had tape recorders that were modified so that a tape could be played back at a series of speeds whose ratios corresponded to the ratios needed to produce all the pitches of a chromatic scale. The title music for my 1971 Christmas Lectures was produced by this technique. The basic tune was produced by recording a single note made by blowing across the neck of a small bottle. This was recorded and then reproduced at all the necessary pitches for the tune by using the multi-speed tape recorder. This tape was then cut up, all the notes required for the tune cut out, cut to the right length and spliced together to make a single tape.

A second component was a percussion accompaniment; a single note was produced by twanging a ruler on the edge of a table and again creating the necessary notes with the variable speed tape recorder. A third tape was made using chords from a harpsichord. The whole was then re-recorded all together and given a little artificial reverberation. The whole process took about three days and the finished piece lasted about 25 seconds!

The skill and patience of those engaged in creating this kind of music were quite incredible and a great deal of background and incidental music for television and films was produced this way in the sixties and early seventies. Many technical developments have occurred that have simplified the various processes, including the multitrack tape recorder that can record sixteen, thirty two, or more tracks on a single tape. But by far the biggest change has been the development of the technique of sampling.

In essence what happens is that a short sound is converted into digital form (see section 5.9) and stored in the memory of a computer. Once this has been done the sound, or fragment of a tune, has become a sequence of numbers which can be processed in all kinds of ways by the computer. It is possible, for example, to play the sounds back at any pitch just as could be done by the variable speed tape recorder—but now it can be done instantly. For example even the relatively cheap kind of keyboard that is sold more as a sophisticated toy than as a musical instrument can record a short sound like a dog bark, or a note made by blowing across the neck of a bottle, and then it can be instantly played back at any pitch from the keyboard. The short recorded sound has become the source of sound quality for that particular keyboard exercise. The quality of the result, as with most digital techniques depends very much on the number of digits used to represent the sound: in other words it depends on the sampling rate; the higher the rate, the better the quality. But, of course, the snag is that as the sampling rate increases the size of memory used to store it and to manipulate it increases extremely rapidly.

The 25 second tune that I mentioned earlier as taking 3 days to create using the cut-and-splice tape technique, could be produced in about 25 seconds plus the time taken to record each sound, multiplied by the number of sounds used. In this case a total of about 2 minutes!

Sampling has become one of the most powerful strands in the production of electronic music. But of course there are problems. I shall mention only three. The first is that a note played back, say, two octaves higher than the original will last only a quarter of the time of the original note and this needs to be taken into account. The second is that the quality of the sound changes when it is played back at a different pitch. For example, in the lectures a boy from the audience was asked to sing a single note and then almost immediately it was possible to play back the note at the pitch of any note on the keyboard of the synthesizer that was used to control the sampler via MIDI. But, for example, when played back in the bass register, the boy's voice did not sound remotely like a bass singer. The sampler used for this and other demonstrations was the S-550, kindly loaned by Roland.

The third, and probably most controversial problem, is related to the question of copyright. It is now perfectly possible to create a piece of music entirely by modifying someone else's recording in all kinds of ways and mixing the results together so that they become almost unrecognisable. But whose is then the copyright? It is beyond the scope of this book to enter into the extraordinarily complicated world of musical and recording copyright. I merely point out that sampling is fine if you create your own sounds to be sampled but can lead to immense problems if samples are taken without permission from other people's work.

Many current synthesizers incorporate banks of sampled sounds that can be used alone or in conjunction with electronic sounds produced in other ways and it is undoubtedly a technique that is here to stay.

5.8 DIGITAL TECHNIQUES

Before discussing digital methods of music synthesis it is necessary to consider the enormous transformation that has followed the introduction of digital techniques into all aspects of sound transmission and recording. The whole process depends essentially on the possibility of encoding all the details of a sound wave in the form of a sequence of numbers which can then be transmitted, recorded, manipulated, etc, and then finally reconverted into sound again. Why should this result in such profound advantages? For those who have not met digital techniques before it may be worth using a very simple illustration.

Suppose that we are trying to communicate a message using a very poor telephone line on which there are all sorts of background noises. The distortion of the message may be considerable. The ancient music hall joke about the message transmitted by word of mouth from soldier to soldier which began as 'Send reinforcements, we are going to advance' and arrived at the other end of the line as 'Send three and fourpence we are going to a dance' is not too exaggerated. But if the message had been sent in Morse code using a note of constant pitch to send the dots and dashes, the background noise would have far less effect and it is likely that the message would have survived unmodified. Sending the actual speech is the equivalent of an analogue technique and sending the Morse code is the equivalent of a digital technique.

The actual technique of encoding signals for digital processing can be understood by using the wave model that we used in section 1.16. In figure 5.7 the sine wave could be represented as a sequence of numbers which are the lengths of the successive pins. These numbers could be transmitted by many different possible techniques and then the wave recreated from the numbers. The accuracy of the reproduction can be increased if the pins are closer together. The accuracy will also depend on how precisely the lengths of the pins can be transmitted. In this actual example the amplitude of the wave is 2 cm. Clearly if the measurement is to the nearest centimetre the representation of the wave will be very inexact; if the measurement is to the nearest millimetre then much greater precision can be achieved.

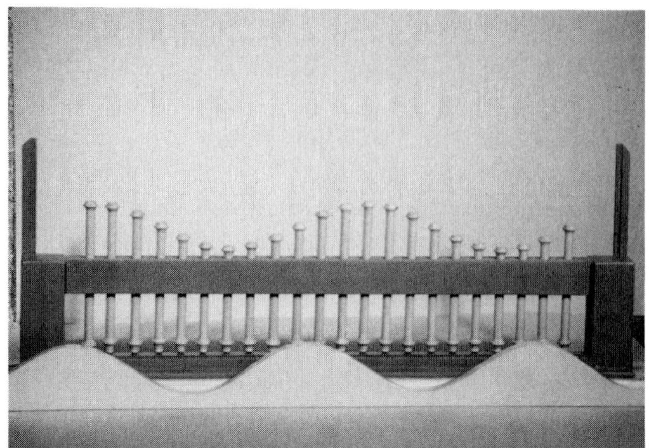

Figure 5.7 In this wave model a sequence of numbers representing the lengths of successive pins could be thought of as digital representation of a sine wave.

In recording or transmitting music digitally the rate of sampling (corresponding to the spacing of the pins) may be upwards of 40,000 per second. The amplitude of the wave at these sampling points is converted into a binary number representing the length. The precision will depend on how many binary digits are used. A binary digit (which may be 1 or 0) is usually referred to as a 'bit' in computer jargon and so, for example, a 6-bit number could be any whole number between 0 and 111111 (which is 63 in decimal notation). If greater precision is needed a larger number of bits is required. For example 12-bit numbers could represent any whole number between 0 and 111111111111 (which is 4095 in decimal notation) which means that the displacement of the wave at any point could be represented with an accuracy of one part in 4096 as opposed to one part in 64 for the 6-bit number.

Thus a sound wave, whether representing a pure tone or the sound of a complete orchestra, could be represented with reasonable precision by a series of 12-bit binary numbers at intervals of not more than 1/40,000 second. This is quite a large amount of information and it is not surprising that it is only since the advent of computer systems with a really large store capacity that digital processing has become practicable.

The first digital system in the field of music to make any impact on the general public was the compact disc. In a compact disc player the plastic disc carrying the recorded information is scanned by a laser beam and, depending on whether the beam hits the unmodified reflecting surface, or a pit, the reflected beam will record either the equivalent of a '0' or a '1'. In the first discs to become available the original music was recorded on tape as an analogue signal which was then converted to digital form for recording on the compact disc. In more recent discs the original recording is done with a digital recorder. Digital tape recorders are only just coming on to the market to the general public. It also seems that compact disc systems that will permit personal recording and erasure to be made will shortly be available.

At this point it is interesting to note that research carried out primarily for the music industry has resulted in progress in completely different fields. For example the miniature laser that is used to scan the compact disc in a domestic player was the result of many millions of pounds expended in research. The size of the market for compact disc players made this worthwhile. But this research has led to the development of much higher power lasers of the same physical size which are now being used for the medical treatment of eye problems. Before these lasers became available the apparatus weighed a very great deal, was about the size of a large office desk needed considerable power supplies and water cooling, and its cost was well out of reach of third world countries. Now that these small lasers are available the size of the apparatus is little more than that of a scientific microscope, it can run from batteries and its cost is down by a factor of about 50.

5.9 COMPUTER SYNTHESIS

In order to synthesise a piece of music by computer it is necessary, as has already been pointed out, to produce a succession of pulses, at least one every 1/40,000 of a second and to be able to modulate the size of each successive pulse to represent the displacement in a sound wave at that particular moment. In principle this is not a difficult feat to accomplish. But it is extremely time consuming and clumsy for any but the simplest of sounds. Two key devices in the development of computer music were the ADC and the DAC. The ADC is the analogue-to-digital converter which takes in an analogue sound—such as, for example, a note of particular pitch, quality, amplitude, and duration—and converts it into a set of pulses representing the same information. The DAC is the converse; it takes in the sequence of digits representing a musical sound and converts it into an analogue waveform that can be fed straight to an amplifier and loudspeaker, or to a tape recorder. In the very early days it was quite possible for even a large computer to take several hours to create a few seconds of music.

The earliest really successful digital synthesis began with the work of Max Matthews, John Pierce and others at the Bell Telephone Laboratories in the USA in the early 1960s. A computer programme called 'Music V' was produced and has become, in many ways, the key to successful music production, though the rate and scale of developments from it have been breath taking.

In essence the idea was to create software routines which would behave rather like the separate units in an analogue synthesizer and then the computer would organise and combine the resulting signals to produce the required end product. The particular units used in Music V are first a group that are the equivalent of oscillators. A single number fed in determines the frequency; a second number determines the amplitude and the resulting output is then available for use. There can, in effect, be large numbers of oscillators and they do not necessarily all produce pure tones; they can produce sawtooth waveforms, square waves, pulses and many other repetitive waveforms.

Secondly there are adding devices that can add together the outputs of two oscillator units for example to add partials of different frequencies, or to add a low-frequency, small-amplitude sine wave to the input determining the amplitude, and thus to give a vibrato effect.

Thirdly there are multipliers which can be used in various ways. Thus the frequency being generated by a whole group of oscillators representing partials could all be multiplied by the same number, so producing the effect of changing key. Or the amplitudes of a group could be changed in the same way, thus producing variations in volume. Finally there has to be a device to collect together and store in sequence all the components of a piece of music. This could be done in the normal computer memory, or more probably if a large number of components is involved, on a disk or on a magnetic tape. The relationship between this kind of synthesis and analogue synthesis can be seen quite clearly from this description. But it is important to remember that the various devices described are only computer simulations of the devices used in an analogue synthesizer and have no separate existence outside the computer programme. It is also important to remember that Music V and the whole family of successors are designed to work on computers that are primarily designed for other purposes, such as scientific or business calculations, etc. The only addition needed is the DAC to convert the final output into sound.

In the 1990s all four of the various lines of development under which I have planned the discussion (analogue, *musique concrète*, digital and sampling; see the beginning of section 5.5) have become diffused and it is now very difficult to describe any one kind of synthesizer without referring to all four historic lines.

5.10 DIGITAL SYNTHESIZERS

I was fortunate in being able to borrow a Yamaha DX7IID synthesizer for the whole period of the lectures and, although a detailed description of its operation would be out of place here (and, I have to confess, somewhat beyond my capabilities!) I shall use an outline of its operation as a means of illustrating the kind of complexities that now exist in modern synthesizers.

(Although having borrowed this as the latest model in December 1989, by the time I was repeating parts of the lectures in Tokyo in October 1990 the machine I was lent was a completely redesigned synthesizer—the SY77, which uses a standard 3.5″ computer microdiskette in place of the earlier cartridge store. This is yet another illustration of the astonishing rate of change in this field.)

The system of the DX7IID is described as 'digital FM synthesis'. We have already discussed the meaning of 'digital' in many different contexts and in this case it means that the keyboard is built round a dedicated computer which simulates all the devices necessary for the synthesis, keeps control of all that is going on and eventually produces an output that can be fed via a DAC to an amplifier and loudspeakers or to tape recorders.

'FM' stands for 'frequency modulation' and, although it is technically related to FM radio, is used for entirely different reasons. The essence of frequency modulation is that a sine wave has its frequency changed at a regular rate. Thus a note of frequency 440 Hz might have its frequency changed, or modulated, by another wave of small amplitude and frequency, say 7 Hz, and the result would be that the pitch of the original note would wobble slightly up and down at a rate of 7 Hz. This would be quite a pleasant vibrato if the amplitude of the modulation is not too high. However, in FM synthesis, which was invented by John Chowning in the late 1960s, the modulating wave has the same frequency as the original wave. The extraordinary result is that, as the amplitude of the modulating wave increases the original sine wave becomes peaked (see figure 5.8) and the number and magnitude of all the harmonic partials increases. Thus, by merely increasing the amplitude of the modulating wave form, a pure tone can be converted into a tone that is increasingly rich in harmonics. A demonstration experiment was set up on the DX7IID in which the modulation amplitude could be controlled with a wheel and so harmonics could be added or taken away smoothly; this together with a display on a real-time frequency analyser gives a very good demonstration of the system, and is also useful to demonstrate the relationship between tone quality and harmonic content. Various other kinds of frequency modulation are also used, for example, instead of modulating at the *same* frequency, a frequency that is an integral multiple of the basic frequency, or even a frequency whose ratio to the basic frequency is non-integral may be used.

As has already been said this is a digital synthesizer, and the discussion about oscillators and modulators should, of course, be read as a discussion of computer simulations of these units. The unique feature of the synthesizer is the way in which the units are all grouped in various ways with units connected in parallel, in series, with feedback loops, etc. The result is an extraordinarily flexible system that can simulate a wide range of orchestral instruments very effectively, but can also produce a marvellous collection of new sounds that could not be produced by any mechanical instrument

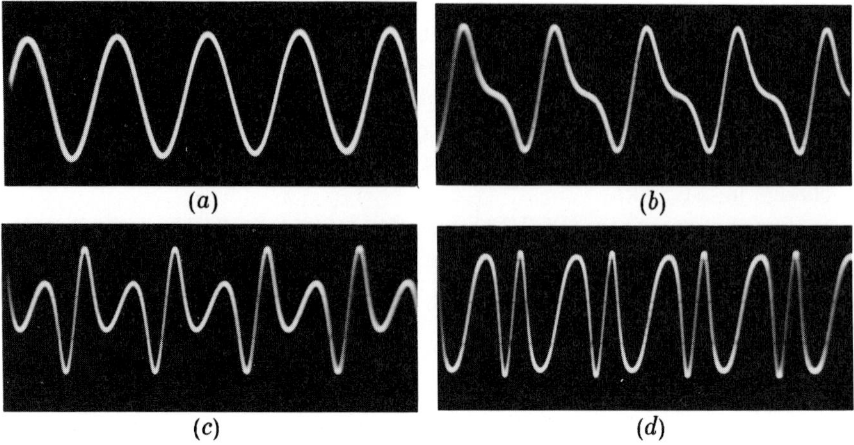

Figure 5.8 Development of a frequency modulated signal: (*a*) is the basic sine-wave carrier; (*b*), (*c*), and (*d*) show the effect of successive increases in amplitude of the modulating signal, which is a sine wave of the same frequency as the basic signal.

(see section 5.14 for a discussion of the implications of this). Once set up, any of the 'voices' can be called up to order by merely pressing one or two buttons. Unfortunately, it has to be said, that the actual setting up, or programming of a new voice is not easy and demands considerable skill. However, the storage capacity is very large and can be increased indefinitely with plug-in cartridges, or computer discs, so this is not a serious limitation in concert use.

Another of the features of this, and of many other synthesizers which is particularly useful for demonstrations of the physics of musical instruments is the possibility of combining two voices. For example, the Yamaha DX7IID was programmed to imitate the steady state sound of several different organ pipes. (8 foot, 4 foot, and 16 foot flute). The same three pipe tones were then synthesised complete with a transient. By merely pressing one button the transient sound could be added. When the sounds of these three pipes were added together, including the three transients, a very convincing imitation of a chamber organ could be produced. In a second demonstration, two voices could be added in variable proportions. For example, one voice could be the steady-state part of the sound of a particular instrument and the other could be the initial transient. The aural effect of gradually introducing the transient is a good way of underlining its importance. The Roland D-50 synthesizer, for example, has an imitation of a steam calliope (or fairground organ) in which the 'breathy' transient and the flute-like steady tones can be mixed in varying proportions; this was a particularly convincing demonstration.

5.11 THE CONCEPT OF MIDI

In section 5.6, on analogue synthesis, I mentioned that one of the major advances was the use of voltage-controlled devices. This led directly to the idea of the sequencer, which recorded voltages in sequence and used them to call up notes of the appropriate pitch, duration, timbre, etc, in order to produce a piece of music. Each sequencer had to be designed for its own particular kind of synthesizer and there were many other limitations. The major one, however, was the one which constantly rears its head in almost any electronic system—that of standardisation.

The next major step forward became possible when practically all synthesizers, no matter what technique was being used for creating the waveforms, began to incorporate their own microprocessor chips that would take on the job of organising and coordinating the various functions of the synthesizer. Then in 1981 at a conference of the American Audio Engineering Society, a revolutionary paper was presented by Dave Smith, suggesting that, if all synthesizer manufacturers could be persuaded to incorporate a standardised unit that could take conventional signals presented in a standard language and convert them to control their own built-in microprocessor then the problem of standardisation might be overcome. Amazingly the idea was actually taken up and by about 1984 every large synthesizer manufacturer in the world incorporated such a device in all its instruments. An enormous amount of argument went on in trying to achieve this standardisation and there remain a few minor differences between different versions.

The core of the system is the 'interface'. That is the digital device that accepts the messages coming into the instrument in digital form and translates them into the appropriate signals to control the instrument; it also passes on messages to other instruments in a chain. The device and the language used are called MIDI, which stands for Musical Instrument Digital Interface. In order to avoid confusion between MIDI signals and audio signals, special 5-pin plugs are used. Any musical instrument operating with MIDI will have three 5-pin sockets, usually at the back. The first is labelled 'MIDI in' and, as its name suggests, is the port through which MIDI data can be fed into the instrument. The second is labelled 'MIDI through' and it simply passes on the incoming signals to any other MIDI instrument without changing them. The third is labelled 'MIDI out' and is the port from which MIDI signals generated by the instrument itself can be passed on to other instruments.

What sort of information can be sent using the MIDI system? Obviously, to begin with, there are the usual data about pitch, duration, timbre, etc. But it can also tell other machines for example, if it is a machine with touch sensitive keys, exactly how slowly or rapidly a key is being depressed. It can pass on data about all the various controls such as volume, pitch variation (usually called 'pitch bend' on modern keyboards) and a great many other

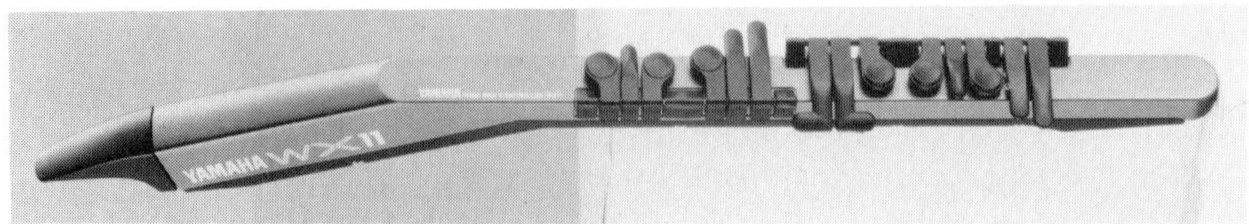

Figure 5.9 A Yamaha 'wind controller' which permits a synthesizer to be controlled by the lips and breath of a player via MIDI. (Photograph by courtesy of Yamaha–Kemble Music (UK) Ltd.)

pieces of information. When MIDI was started it could, in fact, transmit 31,250 pieces of digital information every second. Although it is not yet ten years old MIDI has already seen enormous changes and developments.

To round off this discussion of MIDI I want to describe three of the experiments we performed during the lectures.

The first involves simply linking one keyboard to another via the MIDI cables so that when a piece is played on one, the other slavishly copies the same piece. Of course by appropriate settings of the timbre on each, although they still play the same tune, it can have a completely different tone colour.

The second involves a device called a 'wind controller'. Yamaha kindly lent one to me. It looks rather like a formalised clarinet (see figure 5.9). The mouthpiece looks rather like a clarinet mouthpiece and contains sensors which are able to detect the pressure of the lips and the volume of air being blown into it. By means of the MIDI system the signals derived from these sensors can be made to control clarinet-like (or indeed that of any other wind instrument such as oboe, saxophone, trumpet, etc) sound produced by the DX7IID synthesizer in an extremely realistic way. A simple set of keys control the pitches, and several thumb keys permit the register to be changed instantly.

The third experiment, which was performed largely for fun involved the Yamaha Disklavier (see section 5.13). A piano piece was played by a member of the audience and then the automatic play back system was used to repeat it. By means of the appropriate MIDI connections the same piece was played with an organ-like sound on one of the synthesizers and could be faded in to give a piano and organ duet.

Perhaps the most striking of the many developments was the realisation that ordinary computers could be 'taught' to understand MIDI! It thus became possible to set up a large number of separate analogue synthesizers, digital synthesizers, samplers, artificial reverberation units, multi-track tape recorders and many others all in one studio and to control all of them from a single small computer.

This is in fact precisely what was done in one of the newest of the studios of the BBC Radiophonic Workshop. In this case there is one keyboard, which is the same size as that of a grand piano, and an Apple Macintosh computer which controls everything.

Figure 5.10 shows photographs of an even later studio than the one we filmed for the lectures. The speed of development in music synthesis has already been commented on but in this case the whole studio is run by an Apple Macintosh IIX Computer running on the new System 7, which was only launched in May 1991. It incorporates an impressive array of the very latest software, some of which was devised by the Radiophonic workshop itself and others which were bought in. It includes four synthesizers, a sampler, analogue, digital and magneto-optical recording machines, time controllers, MIDI routing and interface devices, mixing desks and many other features.

5.12 WHY SYNTHESISE ANYWAY?

Every time a new instrument has been devised, whether it was the violin replacing the viol, the tuba replacing the ophicleide, the saxophone moving from the jazz sphere into the classical, or the Spanish guitar moving from the folk style into the classical repertoire in the hands of Segovia, there has always been controversy. But, sooner or later, the shock waves settle down and the new status is accepted by most people. However, the controversy that began with electronic organs and has continued ever since as newer and yet more complex synthesizers follow each other on to the market with great rapidity, is probably of a different order of magnitude.

However, I feel very strongly that a great deal of the problem (as usual) stems from misunderstandings. Many people with whom I have discussed this topic seem to be under the impression that the primary use of synthesizers is to mimic real instruments; some people imagine that electronic organs are really meant to replace real pipe organs. And, of course, some proponents of computer music have been heard to claim that in a few year's time we shall not need symphony orchestras at all.

Let me try to put forward my own view about these various questions and leave you to form your own opinion.

First I will deal with electronic organs. I would contend that, in order to produce an electronic organ that really is indistinguishable in tone from a pipe organ of similar specification and which 'feels' the same to an organist when played, might very well cost more than the corresponding pipe organ. However, that said, it is nevertheless true that an electronic organ costing a tenth or a twentieth as much as a similar pipe organ, could be very acceptable for some purposes. For example, the accompaniment of hymns for a large congregation in church certainly does not demand anything like the quality of tone that is necessary for a concert recital. In other words,

Figure 5.10 Three views of the latest computer-controlled electronic music studio at the BBC Radiophonic Workshop. (Photographer Barry Boxall, Copyright BBC, reproduced by permission.)

the purpose of an electronic organ should not be to imitate a pipe organ, but to produce an acceptable sound in its own right for particular purposes. A pipe organ would clearly be totally unsuited to the playing of pop music in a night club!

The argument about modern synthesizers is perhaps more complicated. I suppose it begins in much the same way as with organs; clearly orchestral music and electronic music are different and each in its own way has an important part to play. But the mere imitation of one by the other seems, at first sight, to be a waste of time. However, closer examination gives a different picture. First, attempting to imitate the sound of a conventional instrument with an electronic synthesizer can be an excellent way of studying the conventional instrument from a scientific point of view. A great deal has been learned, for example about the significance of transients through the process of attempting to imitate them satisfactorily. Secondly synthesizers have greatly extended the range of sounds that are available to composers. The range of instruments available in a conventional orchestra has increased enormously since the sixteenth century and so the quality of sound produced by an orchestra has totally changed. But, though Early Music is more popular than ever, there is no question of its ever eclipsing modern symphony orchestras. In the same way the synthesizer has a firm place both in music designed specifically for it, and also as an additional component in a large orchestra. My own particular taste does not incline to electronic music on its own, but rather in combination with more conventional instruments. That is clearly a matter of personal taste. But, just as it would be a waste of time trying to imitate electronic music with conventional instruments, so is the reverse true.

Thirdly the availability of electronic synthesizers ('keyboards') has made music available in the home. It is possible to buy electronic keyboards for a small fraction of the cost of a piano and perhaps they will fulfil the kind of function that the cottage piano and the harmonium filled in the late nineteenth century. It is common to scoff at the built-in rhythms and accompaniments of some of the keyboards available, but if they begin to foster and develop a sense of pitch and of timing in young people they could very well have a distinct contribution to make in musical education.

5.13 MECHANICAL INSTRUMENTS AND THEIR SUCCESSORS

I do not intend to spend much time on the details of early mechanical instruments except to wonder at the enormous ingenuity that was applied in the design and construction. I suppose the simple musical box is the most junior member of the family. A rotating drum with pins on its surface is placed near to a 'comb' of springy metal strips of varying length. The pins pluck the strips as they pass and so play tunes. Developing from that were various devices in which the set of strips remained a fixture but the drum, or sometimes a large disc could be changed so that a variety of tunes could be played.

The steam organ in a fair ground and the so-called 'player piano' come next. In both, the operation depends on a roll of stiff paper in which holes are punched. The roll travels through the machine and, as it passes over a wind chest, air (or steam) blowing through the holes operates the mechanism of either the organ or the piano. There are still automatic organs in existence in some churches that can be made to play a limited range of hymns if no organist is available.

The most sophisticated of all these is the player piano which is supplied with rolls on which are punched the holes corresponding to a particular pianist's definitive performance of a particular piece of music. But, of course, the interposition of all the mechanical devices between the mechanism and the hammers imposes a characteristic on the music that was certainly not there in the original performance.

In the last few years the application of computer technology has produced a number of different instruments that are direct successors to those just described. I shall pick just one, for the simple reason that it is the one that I was fortunate in being able to borrow for demonstration during the lectures.

It is the 'Disklavier', which was kindly lent to me by Yamaha–Kemble. Both grand and upright versions are available but there would not have been room in the theatre for a grand. The instrument appears at first glance to be a normal upright piano. It has the normal keyboard, pedals, hammers, strings, etc, and can be played exactly as an ordinary piano. Close inspection, however, reveals a small electronic panel in the top right hand corner which has a few switches and lights, a slot for the insertion of a 3.5″ computer microdisk, and a LED display.

The normal piano mechanism has sensors incorporated which detect the movements of the hammers, pedals, etc, and there are also electromagnetic operating devices which will operate the keys and pedals. A piece of music is played in the normal way with the computer mechanism set. The disk can then be reset and the Disklavier will play back the piece (complete with variations in tempo, loudness, etc, and also, of course, any mistakes that were made). The instrument can be operated by means of an infrared remote control and during the replay the speed, key, loudness, etc, can all be modified. It is also possible to record, for example, one part of a duet, or the left hand only, and then to play it back and add the other part to it during the replay. There are many other possibilities, but this description will have given an idea of some of them; its use as a teaching tool is especially valuable.

Although the reproduction of the original player's touch is remarkably good, there is, of course, a characteristic imposed on it which stems from the instrument. MIDI 'in', 'through' and 'out' sockets are provided and the Disklavier can be made to control any other electronic device, however, because of the slight delay occasioned by the electromechanical operating mechanism there is a slight time difference between the 'master' and 'slave'.

5.14 CONCLUSION

Whole books can be written about the extraordinary developments in musical synthesis and, of course, the whole pop music recording industry is so highly commercial that an update is needed almost every week in order to keep up. Inevitably what I have said in this section is only a very small taste of the field. I hope, however, that I have given enough of the ideas to indicate what is possible. I am writing this chapter in the summer of 1991 and I am quite certain that, by the time the book is published, there will have been numerous startling developments that will make much of it out of date.

6

Reflections, Reverberation, and Recitals

6.1 INTRODUCTION

Architecture, like any other subject in which the work of experts is on daily display to the general public, is a fruitful field for controversy, and architectural acoustics is no exception. It is very rare to find any concert hall or music room that meets with the universal approval of performers, listeners and critics. But what is not always realised is that the basic principles, from a scientific point of view, are well understood. The three main sources of difficulty that face anyone involved in acoustic design are first that it is often very difficult to obtain from the client a firm statement of the purposes for which the hall is to be used. A concert hall may be used for anything from oratorio with full choir and orchestra to a solo harpsichord recital; it may also be expected to double as a venue for political meetings, a hall for presenting degrees or other awards, a place for pop music, a cinema, or even an arena for boxing or wrestling. It goes without saying that the acoustic needs of each of these is totally different, and, in any building that is not designed for a single purpose, there must be compromises that may well result in none of the requirements for any one purpose being satisfied.

The second problem may well be closely related. It is that to achieve ideal acoustics for any one purpose may be prohibitively expensive and so again compromises have to be made.

Thirdly, even when an agreed design has been settled with the clients, the acoustic consultants and the accountants, there can be many a slip between the drawings and the finished building. I can give two examples from my own experience to illustrate this point. The first was an auditorium in which a specific requirement of the client, not unreasonably, was that the air-conditioning system should be completely silent. Great effort was put into the design in order to achieve this and it was agreed that I should visit the site weekly during construction, accompanied by representatives of the client, the architect and the builders in order to ensure that the requirements of the design were being achieved. Imagine our horror on one occasion when we found that the large air ducts, which were to

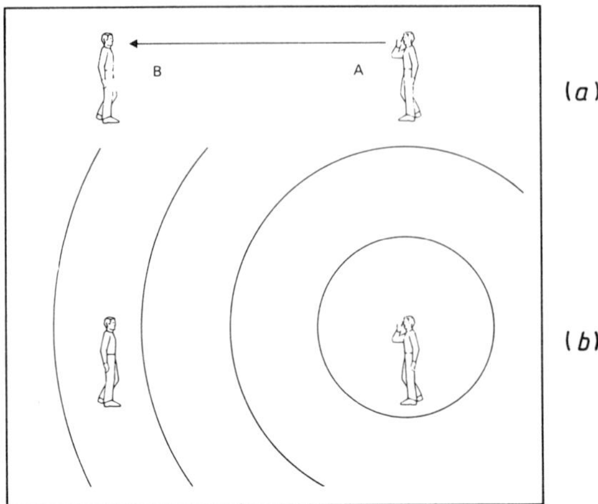

Figure 6.1 (*a*) Conversation up in the air, and (*b*) some of the waves involved.

be suspended on spring supports to minimise noise conduction, had been lowered to rest on the lower bar of the roof trusses, as one of the builders remarked 'to give them more support'. Fortunately this and other problems were spotted in time and the desired end result was achieved. My second example was less happy in outcome, though (for reasons that I am glad to say had nothing to do with my contribution) the building has now been demolished. It was a lecture theatre in which certain areas were to be given a very hard reflective finish and other areas were to be absorbent. Unfortunately I did not see the room until all was complete, including decorations. Fortunately the room was sufficiently small that the fact that the builders had interchanged the absorbent and reflective areas did not have a totally disastrous effect—except on my feelings every time I went in to the room!

6.2 EVERYBODY MUST BE SOMEWHERE

This title was used for the opening section of the chapter on acoustics in the record of my 1971 Christmas Lectures (*Sounds of Music*) and it may strike you as a very obvious statement. The point of it is that it is virtually impossible to consider the behaviour of the sound waves emitted from any source without taking into account the precise surroundings of both source and listener. To emphasise this point I shall begin exactly as I did in 1971 by considering a series of purely imaginary experiments.

In figure 6.1(*a*) two people are shown suspended (perhaps by will power) in the atmosphere, but completely removed from any other solid objects. In these circumstances it would be very difficult to hold a normal conversation

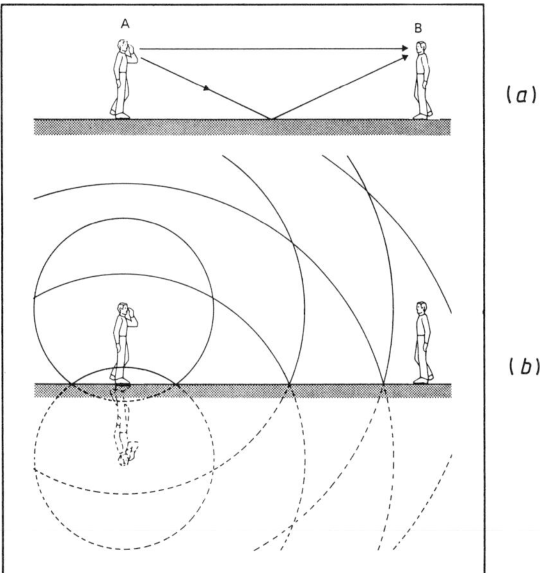

Figure 6.2 (a) Reflection from the ground, and (b) some of the waves involved.

if they were more than about ten metres apart. In figure 6.1(b) the crests of pressure waves created by A as he tries to call to B have been drawn in. (Of course this should be a three-dimensional figure and the crests should be spherical surfaces.) In section 1.2 we discussed waves in general and made the point that a wave is a means of transmitting energy from point to point. Man A is obviously not being very effective in transmitting energy to B and most of the energy is being dispersed in other directions. Suppose A is ten metres from B. Then the energy produced by A will be dispersed over the surface of a sphere ($4\pi r^2$) of radius ten metres by the time the sound reaches B; that is 1257 square metres. But, even if man B has an enormous ear that will collect sound over an area of 2500 square millimetres, he will be collecting only 2500/1,257,000,000 of the energy; that is about one part in half a million of the original energy. So clearly this is an extremely inefficient way of transmitting energy.

But, of course, this is a most unlikely situation and, as the title of the section says 'everybody must be somewhere'. At the very least we usually have our feet on the ground. In figure 6.2(a) the effect of the ground is shown. Man B now receives not only the direct waves, but also those reflected from the ground. In figure 6.2(b) the reflected waves can be seen and the effect is as though there is a second man standing upside down below the first. So far we have assumed that the waves are reflected in the same way that light waves would be from a mirror, but, obviously, this would be true for sound only if the surface of the ground were completely

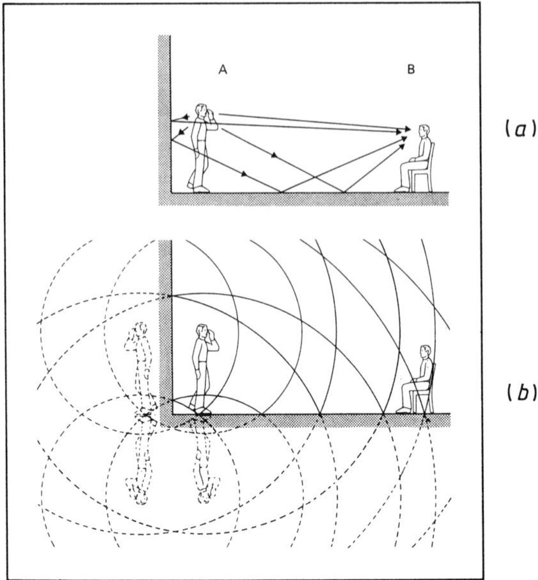

Figure 6.3 (*a*) The effect of a wall, and (*b*) some of the waves and images involved.

flat and extremely hard so that no energy is absorbed. This is never the case, but it nevertheless gives us a way of thinking about the problem which works provided that we do not push the parallel too far. Suppose we now place man A with his back to a wall. There are now four different paths which the sound can use as shown in figures 6.3(*a*) and (*b*). There may be other good reasons why a street orator may wish to have his back to the wall, but it certainly makes good acoustic sense if the audience are to hear well.

Adding another flat surface, a wall or a ceiling, parallel to either of the first two makes a very significant change in the situation. In principle, if the walls were perfectly reflecting the waves could go on bouncing back and forth many many times as shown in figure 6.4. Finally if we complete a room surrounding the speaker and listener (figure 6.5) we have a closed system and, if the wall surfaces were all perfect reflectors, the sound would go on for ever.

6.3 LOUDNESS VERSUS INTELLIGIBILITY

We started putting surfaces round the communicators to increase the efficiency of their energy transfer. But now we have succeeded to such an extent that communication would be impossible. The first sound created would go on indefinitely and each successive sound would add to the first

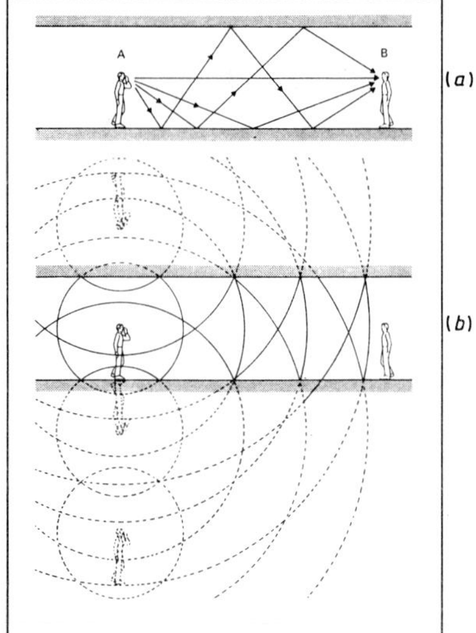

Figure 6.4. (*a*) Parallel floor and ceiling, and (*b*) some of the waves and images involved.

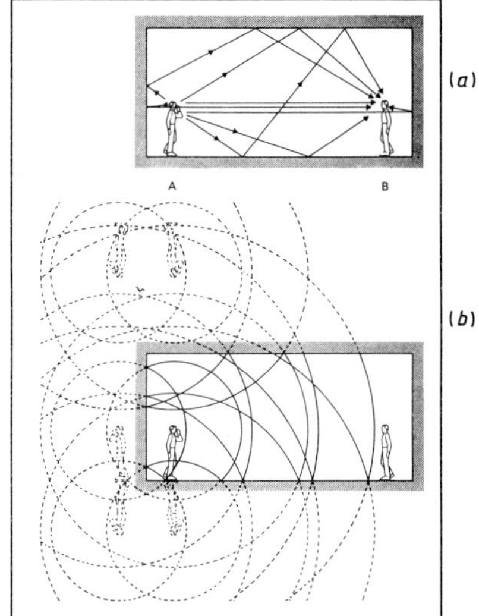

Figure 6.5. (*a*) The complete room, and (*b*) some of the waves and images involved.

until the sound level would be intolerable and each sound would be drawn out to overlap the rest. The intelligibility of the communication would be reduced to zero. An indoor swimming pool with hard tiled surfaces is the nearest approximation that we are likely to find. The noise in such a pool with even a small number of children present makes the point. In practice absorption will always occur to a greater or lesser extent, and the first requirement in acoustic design is to achieve a balance between the need

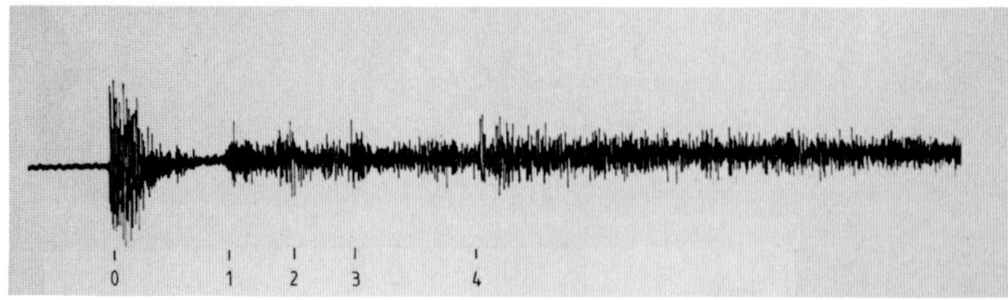

Figure 6.6 Oscilloscope trace produced by a shot fired in a virtually empty indoor swimming pool. The numbers at the bottom indicate the arrival of the first few single echoes; beyond number four there are many multiple reflections.

to deliver sufficient energy from source to listener and the need for good intelligibility.

By careful choice of the placing of the walls and the nature of their surfaces it is possible to arrive at a suitable compromise. But it will already be clear that this compromise is going to depend very much on the exact use that is to be made of the room. A luxuriously furnished room with thick-pile carpets, rich velvet curtains and well padded chairs tends to be a very quiet place. A high proportion of any sound created will be absorbed and you can almost feel the silence 'descend' as you walk in. This would be highly suitable for an intimate chat at close quarters, but totally unsuitable for a large meeting.

The other extreme in most private houses would be the bathroom where most of the surfaces are hard and there is little absorbent. Singing in the bath is pleasurable because the absence of absorption leads to the easy production of a high level of sound and the multiple reflections provide sufficient overlap between successive sounds to blur out wobbles in the breathing and other defects of the voice that would be all too apparent in the sitting room.

The name given to the effect of multiple reflections is 'reverberation'. A useful demonstration is to record the sound of a shot being fired in a reverberant room and then to play back the recording at a very much reduced speed. The shot then sounds more like a drawn out peal of thunder and one can begin to hear each separate reflection from the walls in the early stages. After a very short while the echoes occur so rapidly that the separate reflections can no longer be heard. Figure 6.6 shows the oscillograph trace of the recording of a shot being fired in an indoor swimming pool when virtually no people are present.

6.4 THE WORK OF W C SABINE

So far we have not introduced any really scientific principles into the discussion and it will be useful now to recount the delightfully simple exploratory experiments performed by W C Sabine, beginning almost a hundred years ago. Sabine was a physicist at Harvard and, when a new lecture room was built for the University's Fogg Art Museum (in 1895), he was called in to advise on the problem of its poor acoustic properties. (This is an experience that seems to befall many University Professors of Physics, but Sabine was able to turn the request into a basis for a substantial research contribution.) So effective was his study that only five years later, Symphony Hall in Boston, which was built to his acoustic designs, is still considered to be one of the world's great concert halls.

He very quickly realised that the basic problem was the one we have already introduced—the need to effect a compromise between the reflection of sound needed to enable the audience to receive enough energy to hear the speaker, without at the same time producing so much reflection that the sounds would overlap and become unintelligible.

His first observation in the Fogg lecture theatre was that it was much easier both to give, and to hear lectures when there was a full audience. He went on to deduce (correctly) that the biggest single factor in determining the success of the theatre acoustically was the amount and disposition of absorption.

He proposed a quantity that could easily be measured called the 'time of reverberation'. Roughly this is the time taken for a loud sound to die away to nothing. A crude way to measure it approximately is to clap one's hands, or to fire a pistol shot and to measure the time taken for the sound fade completely. In the offending lecture theatre, when empty, it took about five and a half seconds for a loud sound to die away. It so happened that a neighbouring theatre had long movable cushions, all of the same width, which turned out to be quite effective as sound absorbers. He started to bring cushions in to the hall being studied and made a number of measurements of the time of reverberation and of the length of additional cushions that had been brought in. He used a large organ pipe and measured the time for the sound to die away from the moment the pipe was turned off. He then calibrated the absorption of curtains, carpets, people, etc, by measuring the length of cushion needed to bring the time of reverberation back to its original value after the objects had been removed.

He realised that the product of the amount of absorption present and the time of reverberation for a room was constant. The next step was to find a way of standardising the measurement of the absorption. Obviously lengths of Harvard cushions were not adequate as universal standards! He then realised that the most perfect of all absorbers is an open window; all the sound that falls on it disappears. So it was possible using the same techniques as before to calibrate lengths of cushion in terms of area of open window. Indeed 'o.w.u.s', or open window units, are still used as measures of absorption either in square feet or square metres.

Figure 6.7 depicts a curve drawn from Sabine's results; the hyperbolic

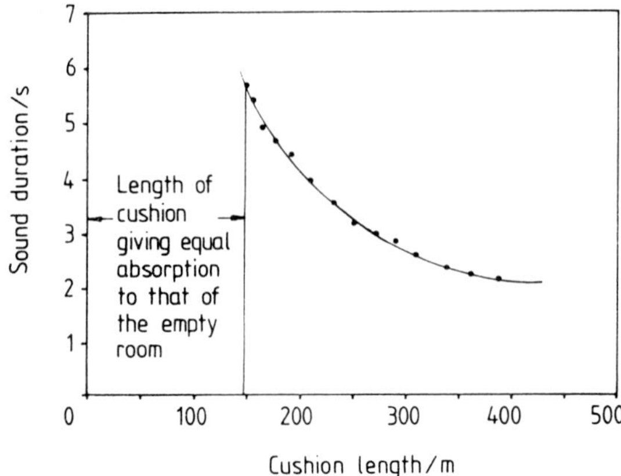

Figure 6.7 Curve drawn from Sabine's results for the lecture room at Harvard; the cushion equivalent of the empty room was calculated using its decay time and Sabine's equation.

shape of the curve, which is the mathematical expression of the constancy of the product of reverberation time and absorption, is clear.

6.5 WHAT TIME OF REVERBERATION IS DESIRABLE?

For a given room, such a graph can be used to predict the reverberation time for a given amount of absorption present. But, though this kind of measurement can be used for predicting the time of reverberation of a room, we need to know what is the desirable time of reverberation to be achieved. The first step is to have a more precise definition of the time of reverberation. Technically it is now defined as the time taken for the energy density of a sound to fall to the level of the threshold of hearing from a level 60 dB higher. (See section 1.11 for explanations of the threshold of hearing and measurement in dB.) In practice the difference between this and our simpler definition becomes important only when the most precise methods of measurement are used.

The basic formula that Sabine devised was

$$(a + x)T = kV$$

in which a is the amount of absorption present in the empty room, and x is the amount added, both measured in open window units (square metres), T is the time of reverberation in seconds, V is the volume of the room in cubic metres, and k is a constant equal to 0.171. (If the volume is in cubic feet and the o.w.u.s in square feet then the value of k is 0.052.) The surprising

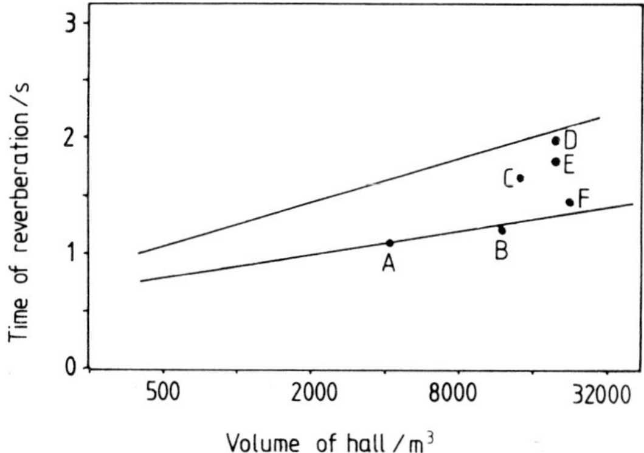

Figure 6.8 Desirable values of T as a function of room volume. The upper line is for music and the lower for speech. It must be emphasised that the choice is very subjective and these values are only a guide. The full circles represent mid-frequency reverberation times for halls as follows: A, the small chamber music room of the Mastersingers building at Nuremberg; B, La Scala, Milan; C, Liverpool's Philharmonic Hall; D, The Concertgebouw, Amsterdam; E, Symphony Hall, Boston; F, London's Royal Festival Hall, before the assisted resonance system was installed.

thing is that the constant k has a value that works for an amazing variety of sizes and shapes of rooms.

Of course the desirable value of T will depend a great deal on the use to which the room is to be put. For music it turns out to be better to have a larger value of T than is acceptable for speech. In figure 6.8 the two straight lines show desirable values as a function of the volume of the hall, the upper line for music and the lower for speech. Of course these can only be taken as guides. The points marked indicate the actual values of T for a number of halls.

So far we have considered a very simplistic view of the problem, though it is surprising how helpful even these approximate ideas can be. The desirable value of T is clearly a matter of subjective judgment and may well be different for the performers and for the audience; it may well be different for sounds of different pitches; and it may depend on the shape of the room. So there are many complex factors to be taken into account.

The time of reverberation could be measured at different frequencies using Sabine's organ pipe method and using pipes of different pitches; the variation can be quite considerable. It also turns out that, though the time can be adjusted by using different materials on the walls (as seat covers, curtains, carpets, etc), each material has an absorption that varies with frequency.

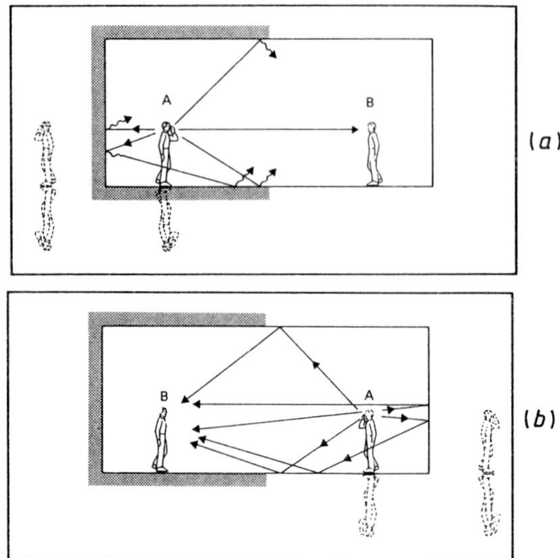

Figure 6.9 A room that is half reverberant and half dead.

6.6 PLACING THE ABSORBENT

As we move nearer to the complexities of real acoustic design we have to consider the effect on the acoustic properties of the disposition of the absorbent material. Here is an imaginary experiment that will help you to see the problem more clearly. In figure 6.9 we see a room that is fairly long and narrow. Exactly half the room is covered with heavily absorbent material and the other half with hard glazed tiles. We will use again the sample people, A and B, used in figures 6.1 and 6.2. First, as in figure 6.9(a), A is singing near the absorbent end and B listens near the tiled end. Then they change places as in figure 6.9(b). Calculation of the reverberation times using Sabine's formula must give the same value of T for both situations since we have not changed the room. But it must be clear that both A and B will experience very different acoustic conditions in the two cases.

In the case of 6.9(a), the images of A in the surrounding walls, floor, etc, will be quite 'fuzzy' and the waves reflected will be quite weak. So A experiences singing in a dead room with little help from reflections. B, on the other hand, will think that the room is very lively and any sound made at that end will reverberate. B will also be puzzled at the low level of sound coming from A.

When the positions are changed A will experience considerable reverberation which will make singing much more enjoyable and B will experience a 'dead room' effect but will hear A's voice very clearly and at quite a high level. Clearly the disposition of the absorbent matters a great deal.

This is, of course, an extreme example, but has in the past found practical use. In the early days of broadcasting when microphones were not directional the configuration of figure 6.9(*b*) was used with B replaced by the microphone. Then the performer (A) had a relatively lively area in which to sing, and in which a good deal of help would be gained from the walls; on the other hand the microphone was in a relatively dead area and so would be less likely to be sensitive to ambient noise.

6.7 PLACING THE REFLECTORS

The distribution of reflection is equally important in designing a room. Consider a large rectangular hall about 30 m long, 15 m wide and 6 m high which is to be used as a concert hall (see figure 6.10(*a*)). The front row of the audience will probably be about 25 m from the back wall and so they will hear the sound reflected from the back wall about 1/6 of a second after the direct sound has passed over (50 m traversed at 300 m per second). The ear–brain system is quite capable of distinguishing two sounds that are 1/6 second apart and so the front row will experience an annoying echo. There will, of course be many other reflections that will all add to the unpleasant effect. Indeed, if the back wall is very hard the front row may hear the so-called 'flutter' effect in which a series of separate reflections is heard in rapid succession, as the successive reflected waves pass over. Clearly absorbent is needed on the back wall. But look again at figure 6.10(*a*); the shaded areas of the wall and ceiling near to the singer are not contributing useful reflections to any part of the audience. The addition of a quite small flare will make this section of the hall much more useful as reinforcement (figure 6.10(*b*)). The same argument can be applied to the side walls as well (figure 6.10).

6.8 SOME UNFORTUNATE CONSEQUENCES OF REFLECTION

So far we have seen that it should not be too difficult to distribute reflecting areas in order to make big improvements in the acoustic behaviour of a hall. But just to show how easily one can fall into traps there is a case on record of the design of an outdoor setting for orchestral concerts that started very logically but ended in disaster. It was decided that, if the orchestra were surrounded by a paraboloidal shell then all the sound would be projected forward rather like the production of a searchlight beam with a paraboloidal mirror. Of course the projection of the sound was a great success. But the shell also acted in the reverse way and focussed every sound made by the audience on to the long suffering orchestra with quite deafening results!

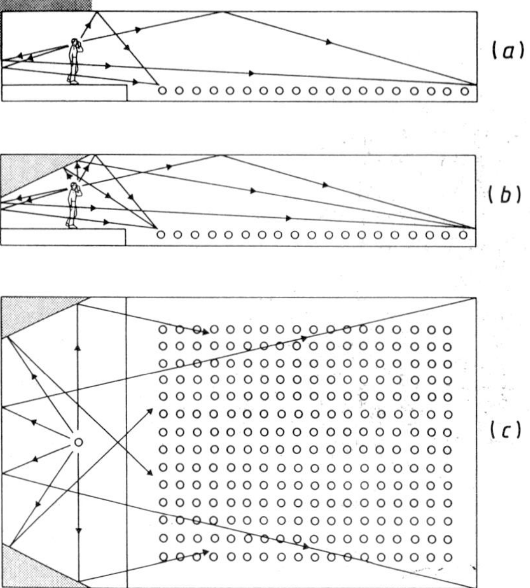

Figure 6.10 In (*a*) the shaded portions of the rear wall and ceiling are not contributing to reinforcement; (*b*) the addition of a small flare makes more use of the reflected sound; and (*c*) the use of side flares perform a similar function.

Curved surfaces generally can be a considerable hazard. Two of the most startling examples, (both of which have long since been cured) are the large reading room of a well known reference library in the North of England and the Royal Albert Hall in London. The reading room was circular in plan and the roof was hemispherical. Unfortunately the height of the roof was chosen to be the same as the radius of the room and of the dome. The desk at which applications for reference material had to be made was exactly at the centre of the room and so was also at the centre of curvature of the dome. Every sound made at the central desk thus radiated outwards with wave surfaces that exactly matched the curvature of the dome at the moment of reflection. The whole of the energy was thus concentrated back on to the central desk, apart from a small fraction that was absorbed. But, of course, it took time for the sound to travel to the roof and back and so the unfortunate person making a request at the desk very quickly became aware that there seemed to be someone else just behind, repeating every word with a delay of about a quarter of a second (the time taken for the sound waves to reach the ceiling and to return). As a result it was extremely difficult to continue a conversation. The effect can be imitated for demonstration purposes by using a tape recorder with a playback head separated from the record head so that speech being fed into a microphone can be fed back into earphones worn by the speaker with

Figure 6.11 Absorbent inverted 'mushrooms' in the ceiling of the Royal Albert Hall which greatly reduced the echo that arose from the focussing effect of the dome.

a delay. It rapidly becomes almost impossible to continue speaking. There were many (mostly apocryphal) stories about events in the reading room. My favourite, which I am quite certain is apocryphal, though the physics is unassailable, concerns students reading at points roughly midway between the centre and the circumference. From this point the sound produced is focussed on a diametrically opposite point. The story goes that a student proposed to a girl sitting opposite him at the same table, and a girl at a table on the opposite side of the central desk accepted him!

This reflection across from one side to the other can lead to strange effects. The sound could rebound several times (I once counted up to eleven transits) and dropping a book or a movement of a chair sounded almost like a burst of machine gun fire. The cure was simple in theory, though awkward to achieve in practice; it was to coat the surface of the dome with absorbent material.

The echo in the Royal Albert Hall was not quite so dramatic but had a very similar origin. In this case the cure was to suspend the familiar absorbent 'mushrooms' from the ceiling (see figure 6.11).

Various London theatres in the past had echoes caused by the reflections of sounds on the stage being focussed by a dome on to the front row of the dress circle. It is said that, in the time of the Inquisition, there were churches in which the confessional was at one focus of a dome and a point in

Figure 6.12 Multiple reflections of a doll placed between two parallel mirrors.

a gallery was at the other! And, of course, in a less sinister situation, some whispering galleries (but not that at St. Paul's Cathedral, where the focus of the dome is far too high and the sound proceeds by repeated reflection round the walls) operate in exactly this way.

The echo with which most people are familiar is the one experienced out of doors, from a large flat wall. In the days just after the Second World War, when there were still many bombed sites that had been cleared but not built on, it was quite common to find an expanse of ground with a high wall at one end and this would give a splendid echo. If the wall were, say, 80 yards away the echo would have a delay of about half a second and would be fairly easy to detect on a quiet day. A different kind of echo is that sometimes heard as one walks down a narrow alleyway between two high buildings. I can remember noticing the ringing sound of my footsteps and being told that the pavement must be hollow. In fact it is due to a multiple echo. A single footstep is reflected many times back and forth between the walls—rather like the infinite sequence of reflections in a pair of parallel mirrors (see figure 6.12) and so the sound heard is a succession of clicks at regular intervals. If the listener is midway between the walls and the walls are 1 metre apart, the succession of reflected images will be 500 mm apart, so if the speed of sound is about 300 m per second, the frequency of the arrival of the pulses will be 600 Hz—which of course is perfectly audible as a musical note.

Before leaving the subject of echoes we should mention the opposite effect; i.e., the total absence of reflection. When microphones are to be calibrated or other acoustic test equipment is undergoing trials it is convenient to use a room that is completely free of wall reflections. Such rooms (described as 'anechoic chambers') form an essential part of the equipment of most acoustical laboratories. The walls, floor, and ceiling are covered with wedges of sound absorbing material and a metal grid floor is provided on which the operator can walk and on which equipment can be set up. The psychological effect of standing in such a chamber is most disturbing because the sense of balance is deprived of some of the signals on which it depends (reflections from walls) and the feeling of isolation from one's surroundings is profound.

6.9 SOME SUBJECTIVE PROBLEMS

We only have space to discuss one or two of the subjective problems that arise and I have chosen the ones that I consider to be most significant. As soon as we walk into a room our ear–brain systems begin to make a series of incredibly complex, and entirely subconscious, calculations based on the reflections from the walls of small sounds like footsteps, or the rustle of clothing. In a very short time we have a very complete picture of our acoustic environment. Blind people are very much aware of this phenomenon and can quickly make judgments about their surroundings. The most obvious example for sighted people is the effect of walking into an anechoic chamber as was mentioned at the end of the last section. The senses record the totally absorbent walls as empty space and yet our eyes tell us that we are in a small room: hence the conflict.

> But how does this affect our experiences in a concert hall? Benade (1977) has given an account of the complex relationship between a player and the room surrounding the instrument. He points out that for every frequency produced there is a system of standing waves set up in the hall and in order to sample the sound efficiently the player tends to move the head, and sometimes the instrument as well in order to move in and out of the null points. Listeners too can move their heads slightly. But the instruments themselves radiate different frequencies in different directions and so the sound field set up becomes very complicated indeed. But the remarkable human brain takes account of all the complications and compensates for them to give us a good impression of the sound being created. However, if a recording is made with a single fixed microphone the variations in sound level at different frequencies will be faithfully reproduced, but the ear–brain system of the listener to the recording is deprived of all the information about the room, etc, avail-

able subconsciously to the listener of the live performance. The recording therefore sounds quite unnatural.

Incidentally a full account of the extraordinary variations in radiation direction of notes of different frequencies from most common instruments is given by Meyer (1978) (already mentioned in relation to the clarinet in section 4.12). For example it may come as a surprise to learn that, whereas between 400 and 600 Hz, a cello radiates most of its sound forward, at 300 Hz and between 800 and 1000 Hz almost all of it goes backwards!

Not so surprising when one considers the way in which the instrument is held, is that practically all the sound from a French horn travels backwards! But I have yet to see a horn player seated with the horn pointing to the audience.

Another subjective phenomenon is called the 'precedence effect'. If the ear–brain system receives the same signal twice with a short time interval between, it will not always detect the two separate sounds. For instance if two clicks, each of very short duration, are made less than about 5 ms apart, then only one click is heard. If the two signals are of music or of speech there can be a delay of 30 to 40 ms before they can be detected as two sounds. This, of course, is a very good thing, otherwise we should be conscious of the arrival of each separate reflection from the walls in a concert halls. Clearly the ear–brain system works with a kind of rule that says, 'If identical sounds are heard in very rapid succession then they are really only meant to be heard once'.

But there is a second point that arises. If sound is falling on the ear from several different directions and being fused into a single sound, nevertheless the brain will recognise the first sound to arrive as being that from the direction of the source. Thus in a hall with relatively hard reflective walls the audience will still sense that all the sound is coming from a singer on the stage even though a high proportion of the sound is arriving from the side.

This is the effect to which the term 'precedence' is, strictly speaking, confined. It becomes particularly important when electronic sound reinforcement systems are in use because if the later sound is very much louder than the earlier one (say 10–15 dB) then the louder one will define the direction from which the sound appears to be coming.

Incidentally it is claimed by some people that the precedence effect plays a part in producing the 'chorus' effect. That is the special quality of fusion produced by a group of singers, or string players singing or playing together which differentiates the sound from that of a single louder instrument or voice. No matter how well trained the group is there are bound to be very slight time delays between the various components and, though the brain fuses them together into a single sound, it is nevertheless 'aware' that the sound is not really from a single source.

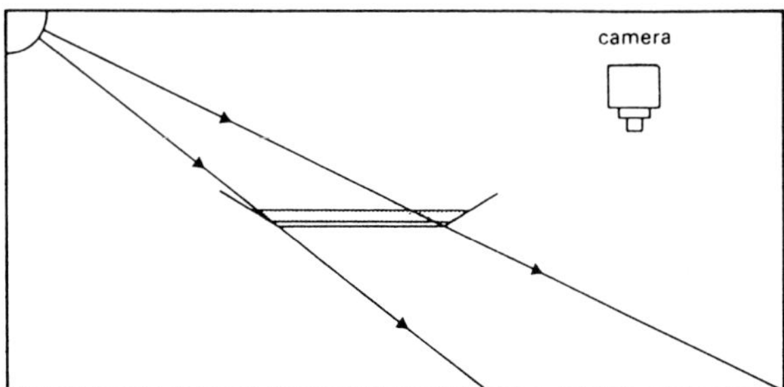

Figure 6.13 The arrangement for casting shadows of the ripples on the surface of a ripple tank.

This has been a very superficial account of the problems facing the acoustical designer but I hope it will at least have given a flavour of the subject and, in particular, have emphasised the importance of going beyond the basic physics into the subjective considerations if the subject is to be properly understood.

6.10 METHODS OF ACOUSTIC DESIGN

Nowadays, of course, three-dimensional drawings by computer can be used together with ray tracing programmes to explore the behaviour of sound waves in a building using optical principles. Reflections are not as perfect as for light by mirrors, but nevertheless a great deal can be learned from such studies. However, some of the methods used in the past are of sufficient interest in themselves to warrant a short discussion.

One of the earliest techniques, and one which is still very useful as a demonstration aid in lectures on acoustics, is the ripple tank. Thomas Young used a ripple tank round about 1800 in his work on the interference of light and essentially the same system can be used in acoustic demonstrations. Figure 6.13 shows the arrangement. Single or multiple ripples can be made on the surface of the water at a point corresponding to a source of sound in a plan or elevation of the hall created in solid wood or metal and placed in the tank. The progress of the ripples is quite easy to follow. Figure 6.14 shows the build up of reflections from two bars in the tank to simulate the hypothetical experiment of figure 6.5(*b*).

One of the problems facing acoustic designers is to decide exactly what a given type of performance is likely to sound like in the finished hall, ideally before it is built. One ingenious approach to this problem was used by the BBC round about 1970 in designing a large studio for orchestral use at the Maida Vale site. The idea was to use a scale model of the hall concerned and to test it out using scaled down sound waves. The model used is at one eighth of the scale of the actual studio. Figure 6.15 shows a view inside the model together with a view of the actual studio. Reducing the scale of the sound waves would seem to be a difficult procedure but, it turns out, that there are several factors that work together to make it a relatively easy task. The trick consists of first recording test sounds in as dead an enclosure as possible. (See the note on anechoic chambers in section 6.8.) This recording is then played back at eight times its normal speed through a small loudspeaker inside the model.

Playing the recording back at 8 times the speed means that all the frequencies are multiplied by eight, which in turn means that the wavelengths are scaled down by a factor of eight. In other words they now match the scale of the model. But the rate at which the sounds are being produced

Figure 6.14 *Opposite.* The build up of reflections from metal bars placed in a ripple tank to simulate the conditions of figure 6.3. (*a*) A sound is initiated, (*b*) the first reflections from the rear wall and floor occur, and (*c*) waves now fill the whole area.

(*b*)

Figure 6.15 (*a*) A one-eighth scale model of the BBC Maida Vale Music studio, and (*b*) the actual studio.

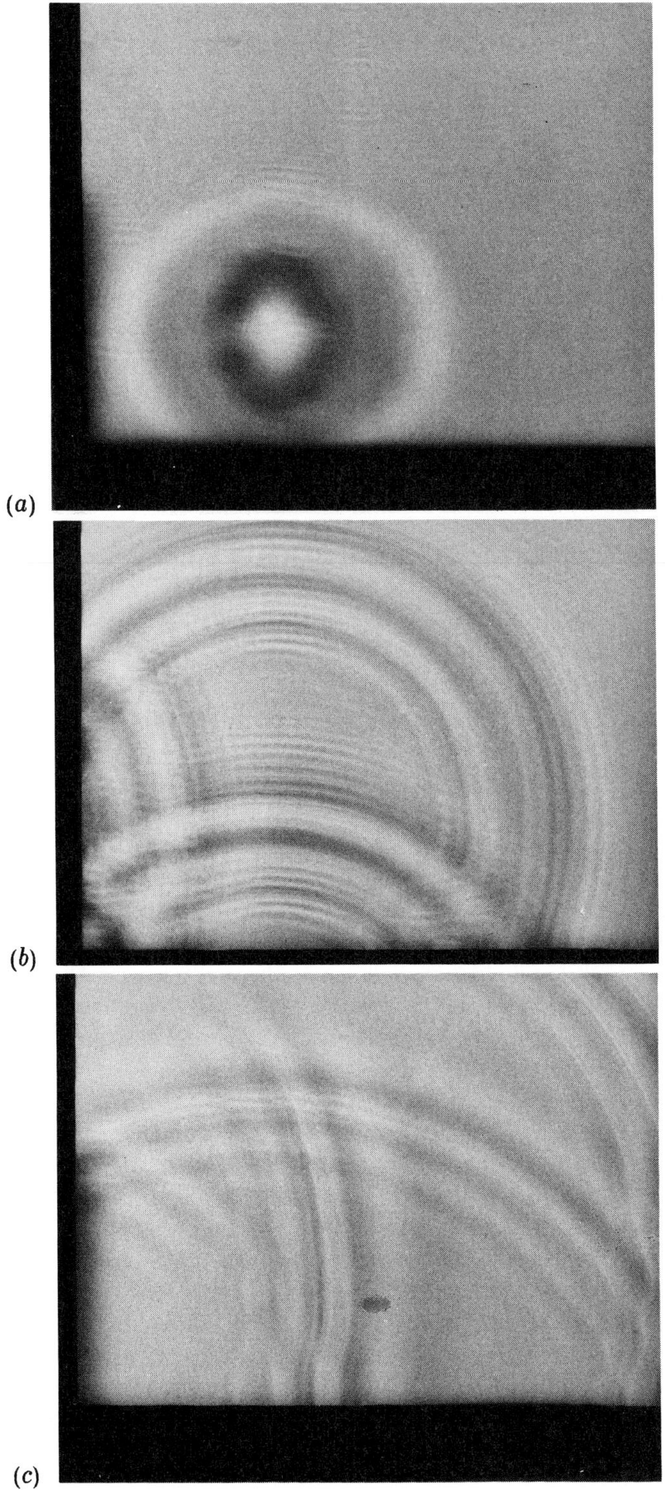
(a)
(b)
(c)

is eight times faster so the time intervals between notes will bear the same ratio to the dimensions of the model as they do in the real hall. The resultant sound in the model is then picked up by a suitable microphone and the results are again recorded. This final tape is then played back at one eighth of its normal speed so that all the frequencies, time intervals, reverberation times, echo delays, etc, that there may be are restored to the normal scale.

The resultant sound is surprisingly close to that in the real hall, but now of course it is very easy (and cheap) to change the positions of reflectors, absorbent material, etc, in the model and to predict the resultant effect in the real environment.

The main problem was to find material that can behave in the model with scaled acoustic properties to represent the materials in the real hall. Polystyrene foam blocks of appropriate shape and size turned out to be a good simulation of the human body to represent members of an audience.

6.11 ADJUSTMENT OF THE ACOUSTICS

There are three kinds of adjustments that can be made to the acoustics of a room after it has been built. The first involves changes in the distribution of absorbing and reflecting surfaces; these are once-and-for-all adjustments that are part of the commissioning process when the hall is complete but not yet handed over. The second involves movable parts of walls or reflectors that can be changed to suit different purposes. The third involves some kind of electronic modification of the acoustic characteristics.

Modern computational methods have made the 'tuning' of a hall after completion far less difficult than it was at one time. Early methods of calculating times of reverberation, etc, were not particularly effective, largely because of the large number of unknown factors like the absorption coefficients of the various materials used, especially during the decoration process. But even 25 years or so ago it was still quite common for relatively large changes in the ratio of absorbent to reflective surfaces to be necessary after completion. I mentioned in section 6.4 that it is not an uncommon experience for professors of physics to be called in to advise on acoustic modifications. In fact as far as I am concerned I first began to learn something of the problems about 40 years ago as a young lecturer when the Principal of my College called me in one day and asked me to 'do something' about the appalling reverberation in his office. Fortunately for me, in more ways than one, the application of some very elementary principles achieved a remarkable improvement. However, the kind of modification that is possible in very large halls is very limited. Also about forty years ago attempts to achieve adjustable acoustics were being made. For example, when the Free Trade Hall in Manchester, the inside of which had been destroyed by enemy action, but the outer shell of which remained intact, was being refurbished as a base for the Halle Orchestra, it was decided to include large panels on the side walls which could be reversed and had absorbent on one side and reflective material on the other side. I am not sure what eventually happened to them, but it was a considerable task to change them over.

In one of the very latest concert halls to be built—Symphony Hall in Birmingham (see section 6.12)—there are large panels of absorbent material that are mounted vertically in slots in the ceiling and can be lowered by means of motors to any given depth and so can make more-or-less continuous changes in the acoustic properties of the room.

Electronic modification has been tried with varying degrees of success. The principle is to site microphones in the room to pick up the sound, then modify the sound in various ways and feed it back into the room using loudspeakers. The late Lord Bowden used to say that in principle it ought to be possible to make a telephone box sound like Symphony Hall in Boston. The practical problem was always that of decoupling the microphones from the loudspeakers. One early attempt was made in a large lecture theatre at the University of Manchester Institute of Science and Technology. Thirty or more loudspeakers were mounted round the rear wall and sides and were fed with sound picked up by microphones placed in front of the instrumentalists. Variable time delays could be introduced in the feed to simulate varying amounts of reverberation. In spite of lengthy tests it was extremely difficult to arrive at a conclusion about whether the audience could notice any difference when the system was turned on. What did appear quite clear, however, was that the players themselves could very definitely distinguish when the system was turned on and they very much appreciated the extra 'liveliness' of an otherwise rather dry hall. This in fact draws attention to the extremely important question of suiting the performers as well as the audience. Indeed it might almost be said that any compromise should be weighted in favour of the performers.

Another system that was tried with rather more success than the purely electronic modification was to allow sound from the auditorium to enter the cavity between the suspended ceiling and the structural ceiling and then place microphones in this space, which was highly reverberant. The sound picked up was then mixed with direct pick up from the theatre itself and the combined result fed to the loudspeakers in the hall.

The method of using a real room as a source of additional reverberation has been used in many buildings. It was used,for example, to improve the reverberation of the hall used for broadcast concerts by the BBC Northern Orchestra. But, of course, in this case, the problem of feedback did not occur because it was only in the broadcast sound that the added reverberant material from microphones placed in the reverberation room, was used. The players in this case did not experience the added reverberance.

One of the most famous examples of electronic assistance to the resonance of a concert hall is that of the Royal Festival Hall in London. The problem was that, while reverberance at the higher frequencies was adequate, at the lower end reinforcement was needed. The problem of feedback was tackled by dividing up the frequency range into relatively narrow bands. For each band there would be a set of modes of vibration of the hall—rather like three-dimensional Chladni figures. A microphone fitted with a filter at a particular frequency could then be placed at an antinode for that frequency and the loudspeaker placed at the node would be decoupled from it. The system was approved by large portions of the audience. But there remained a problem with instrumental soloists. A certain well known harpsichord player disliked it intensely because he said that, as he played, different notes 'came at him' from different directions in the auditorium as different loudspeakers came into play.

6.12 SYMPHONY HALL, BIRMINGHAM

At the time of writing, one of the most recent and most successful concert halls is Symphony Hall in the International Conference Centre in Birmingham, opened by the Queen in June 1991. Since it exemplifies many of the techniques and problems that have been discussed already in this chapter it will be useful to give an account of the completed hall.

It seems to me that one of the most important factors contributing to the success of the hall is that the team of consultants worked together from the very beginning. Indeed the Concert Hall Design Consultants, the Theatre Consultants and the Acoustic Consultants were all part of the integrated team of a single organisation: Artec Consultants Inc.

The hall, though first and foremost designed for classical music concerts with a large orchestra, is also required to act as a hall for organ recitals, choral concerts, solo recitals, rehearsals for all kinds of events, meetings, plenary conference sessions and for many other purposes. The finished result is a remarkable example of how such a hall can have its acoustic properties modified to suit many different occasions without the use of any electronic devices. (By that I mean audio-electronic devices; many electronic and computer controlled devices are used in the process of controlling the physical characteristics of the room.)

The basic design involves a relatively narrow hall with flat, parallel walls which give side reflections that give the sensation of being surrounded by the sound. This effect is still further enhanced by the use of what are described as 'reverse fan walls' at the rear to provide additional reflections all of which have travelled sufficiently short paths that no problems with echoes or time delays can occur. The total volume of the hall itself is 25,000 cubic metres and it seats 2250 people.

The designer, Russell Johnson, has provided a line diagram (figure 6.16)

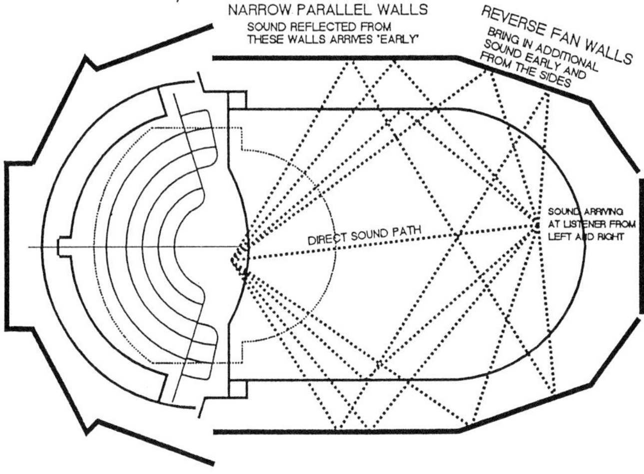

Figure 6.16 Symphony Hall, Birmingham: sound paths between a soloist and a listener at the rear of the stalls.

Figure 6.17 Photograph of Symphony Hall, Birmingham, taken by the design consultant, Russell Johnson.

and a photograph of the hall (figure 6.17) which illustrate many of the points made.

The three factors that seem to predominate in all the descriptions of the sound quality by a wide range of musicians, critics and the public are first the feeling of being surrounded by the sound already mentioned. Secondly the 'warmth' of the sound, which is achieved by paying special attention to reflections of the low frequency components by using relatively massive concrete reflectors; the ceiling and other reflectors are made of 200 mm thick concrete, coated with various materials chosen for their acoustic characteristics as well as for their appearance, And thirdly the clarity of the sound. This is achieved largely through the closeness of all the reflecting surfaces, including the canopy over the orchestra, to the listeners.

But what about the changes that can be made to suit different occasions? First the reflecting canopy over the orchestra position can be raised and lowered over an enormous distance. This has several functions. First it provides the very necessary reflected sound to the members of the orchestra themselves. Again the reflector is near enough to eliminate long time delays and enables the members to hear their own contribution as well as those of the rest of the orchestra so that the right balance is made easier to achieve.

Secondly it can be set very low in order to enhance speech from the stage area and to minimise sound 'lost' in the upper portions of the auditorium.

Thirdly it can control the extent to which sound is admitted to the reverberation chamber.

This reverberation chamber is a U-shaped volume that is wrapped round the end of the hall. Huge concrete doors that can be opened and shut at will are provided which allow the sound from the main hall to reverberate round and then to re-enter the hall. The volume of the chamber is 12,700 cubic metres and thus it is possible, in effect, to increase the reverberating volume of the hall by up to 50%.

There are also large sound absorbing panels that can be lowered into the main hall to cover a high proportion of the wall surfaces.

At the beginning of this chapter I said that though the physics of the behaviour of sound in concert halls is now well understood there were other problems that tended to make the achievement of good acoustics in practice very difficult. It seems to me that Birmingham's Symphony Hall is now a classic example of how these problems may be overcome.

6.13 NOISE IN BUILDINGS

Noise in buildings is a highly controversial topic whether we are talking about the next-door neighbour's Hi-Fi system being played at maximum loudness, the noise of the air conditioning system in a concert hall, or the problems of adjacent musical practice rooms. Just as with the general topic of musical acoustics, the physics is well understood and the problems lie with the conflicting requirements of aesthetic design, cost and the sometimes odd requirements of the organisation that is paying the bills.

Noise between rooms in blocks of flats or between neighbouring houses can be either airborne or structure borne. A surprisingly large amount of airborne noise can be transmitted by badly fitted windows or doors and, generally speaking heavier doors or walls are more effective in reducing noise levels than flimsy ones. If a building has a steel frame then it is important, if possible, to prevent noise getting into the frame in the first place. Once noises get into the structure they can travel long distances with little attenuation. Apart from the frame itself, noise is easily transmitted along water pipes, ventilation ducts, central heating pipes, etc.

The noise usually associated with air conditioning systems arises from either the fans used to drive the air round, or from the noise of air movement. The fans should be mounted on vibration absorbent mounts, but it is also important that there should be insulating joints between the fan exit and the ducting, otherwise the sound will simply be conducted round the building by the metal of the ducts. Air noise can be cut down by using ducts of relatively large cross-section so that low velocities can be used. It is also possible to fit air exits that contain damping elements (rather like the mufflers used to silence car exhausts).

The problem of isolating practice rooms in schools or colleges is an interesting one. It is surprising how many people think that the first step is to insulate the building from exterior noise, such as traffic noise. However, this is not by any means the most important factor. It is far less distracting for someone in a practice room to hear a little traffic noise than to hear someone else practising in the room next door. In fact a little traffic noise can be an advantage provided that it is relatively neutral. It provides a threshold level which tends to mask any sounds leaking from other rooms. The principle is really that of the Weber–Fechner law (section 1.10) and is the argument (scientifically valid) used by countless children to excuse playing records while they do their homework! A relatively neutral piece of background music is far less distracting than the sound of a baby crying or of a domestic dispute.

In fact, in practice rooms where the isolation of each from its neighbours is the ideal goal, I have recommended the use of a small loudspeaker fed with an adjustable level of white noise, which is certainly neutral enough to mask interfering sounds without itself being irritating.

Noise that is immediately recognisable is infinitely more distracting than much louder but unidentifiable noise. I was once asked to comment on some designs for a small hall in which music was to be played. I immediately noticed that there were some toilets proposed on the other side of the wall of the hall itself. I suggested that they should be moved to another wall because I can imagine nothing more distracting than the the noise of a flush in the middle of a pianissimo solo!

The measurement of noise levels is a subject of endless controversy and one of the main problems is that discussed in section 1.17. The human brain is always one step ahead and takes into account so many factors other than the simple physical parameters. Many years ago before the second World War I lived in a house where a railway ran past the bottom of a small garden. Steam engines pulling heavy goods trains regularly went by and visitors used to ask how we could stand the noise. But strangely enough after a few days we never noticed the trains. But the much rarer experience of an aircraft passing overhead was very noticeable!

6.14 CONCLUSION

I am very conscious of the superficial nature of the discussions of this chapter. But the subject is immensely complex and my main purpose was to indicate some of the main problems that have to be dealt with by architects. It seems to me that the most important lesson that has been learned over the last twenty or thirty years is that the various personal and psychological factors that occur in building design are every bit as important as the purely technical or scientific ones. In fact one could argue, as has been discussed in the musical part of the book, that science must be applied in the concert hall, not just in the laboratory, so in architectural acoustics it is not sufficient to make calculations using only the physical parameters of the building.

Appendix A

Holographic Interferometry

A.1 INTRODUCTION

There are many different ways of using the techniques of holographic interferometry in studying the vibrational properties of musical instruments. It has been used to study such widely diverse musical topics as the vibration patterns of bells, the contribution made by the horns of brass instruments such as the trombone, the vibrational properties of the bodies of violins, the coupling between the vibrations of an instrument such as the guitar and the body of a player, and many others.

The fundamental principle remains much the same and is described here. We will first look at the basic principle of holography. Figure A.1 shows the idea in diagrammatic form.

It is important first to remember that light waves belong to the family of electromagnetic waves and behave in many ways like radio waves except that whereas radio waves have frequencies measured in kilo Hertz or even mega Hertz, light waves have frequencies measured in hundreds of millions of mega Hertz. Radio transmitters emit continuous trains of waves that can last for quite long periods of time without interruption. However, light waves normally consist of enormous numbers of short bursts of energy; physicists describe such a light beam as being incoherent. However, since 1960, light sources that are coherent, that is they emit continuous trains of waves in the same way that radio sources do, have been available. They are LASERs (which is an acronym for 'Light Amplification by Stimulated Emission of Radiation').

Laser light is not only coherent, but it is much more intense than usual light sources, leaves the source as a parallel beam without the need for lenses, and is highly monochromatic.

If such a beam falls on an object (as in figure A.1(a)) the light is scattered and we could form an image as with ordinary light by placing a lens at L. If instead of placing a lens at L, we place a photographic plate we would not expect to see anything on development of the plate, other than complete fogging. However, if we direct a portion of the incident beam by a mirror

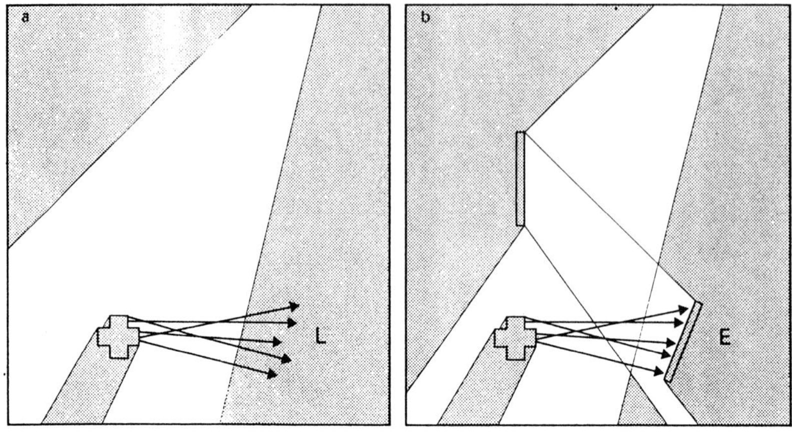

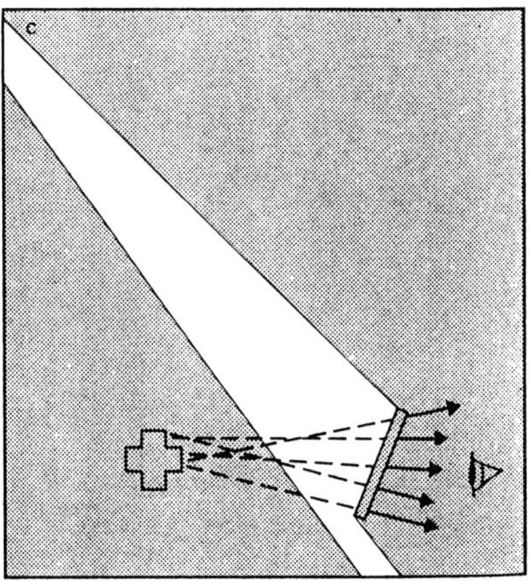

Figure A.1 Principle of the hologram.

so that it lands on the photographic plate (as in figure A.1(*b*)) a remarkable effect occurs that does not occur with ordinary, non-coherent light. Every point on the object doing the scattering, scatters light to all points on the photographic plate, but each individual wave will have travelled a different distance from the object to the plate. The result is that the scattered wavelets will have different phases relative to the beam reflected from the mirror (known as the reference beam). Those that are in step will add up to give light, and those that are out of step will produce no light. Thus a highly complicated and finely detailed interference pattern will be formed on the photographic plate. Notice that changes in path difference of about half a wavelength of light, that is a few tenths of a millionth of a metre, will change a patch from light to dark.

Suppose we now develop the plate and put it in a laser beam (see figure A.1(c)) so that it is in the same relation to the beam as the plate was to the reference beam when the initial hologram was made. The dark and light patches on the developed plate will turn the beam into a set of beams that are exactly like the waves that resulted from the interaction of the scattered waves and the reference beam in the first stage. On looking into the developed plate therefore, our eyes receive a set of waves identical to those we should have received if we had looked from position E in figure A.1(b), and hence we imagine that we can see a full three-dimensional image of the object as it was in figure A.1(b).

This is the normal technique of holography. If we make a hologram in this way of an object that is vibrating, say by being driven with an electromagnetic vibrator system, then we not only see an image of the object, but, it is crossed with light and dark bands, or interference fringes that can give a great deal of information about the nature of the vibrations. This kind of interferometry depends on the fact that when the object vibrates it exists for the greater part of each cycle either near one extreme position or the other and crosses over rapidly from one extreme to the other twice in each cycle. The hologram can be thought of as consisting of two objects (corresponding to the extreme positions) and, since the images are formed by coherent light waves, there will be interference fringes between the two images. The result resembles the sand figures of the Chladni plate, but there are more fringes and, since no sand is involved, there is far less potential damage to the instrument. Examples of the application of the technique to a Chladni plate and to guitar and violin plates are given in sections 3.2 and 3.4.

Appendix B

Pitch and Frequency

B.1 INTRODUCTION

In section 1.5 we discussed some basic relationships between pitch and frequency and, later, in section 5.2 we discussed the setting up of musical scales and, in particular the scale of Just Intonation and the idea of temperament. There have been many systems proposed for the designation of notes and for the precise measurement of pitch and it seemed appropriate that we should bring some of them together so that the relationships between the systems used in this book and those used in other publications could be presented in an accessible form.

B.2 SYSTEMS OF PITCH NOTATION

Figure B.1(a) shows a diagram of the 88 keys on a piano numbered serially from left to right. This is a fairly universal and unambiguous way of designating notes. Figure B.1(b) shows how the notes on the musical stave relate to these numbers; only the seven 'C' notes are indicated. The five lines and four spaces of each of the two staves are the most universal ways of writing down music for performance, though other methods exist, especially in relation to electronic music. However, they are not yet sufficiently standardised to warrant inclusion in a book at this level of generality.

The single line midway between the staves is known as middle C and is note No 40 on the piano keyboard.

There are eight or nine systems of so-called 'staveless' notation in use and when reading any textbook it is advisable to check carefully which system is being used. The four most commonly used are: Helmholtz, American standard, American Alternative, and English Organ builder's. In my earlier book *Sounds of Music* I used the American Alternative system. But the American Standard seems to be accepted more and more and so in this book I have used that system. The relationships are given below.

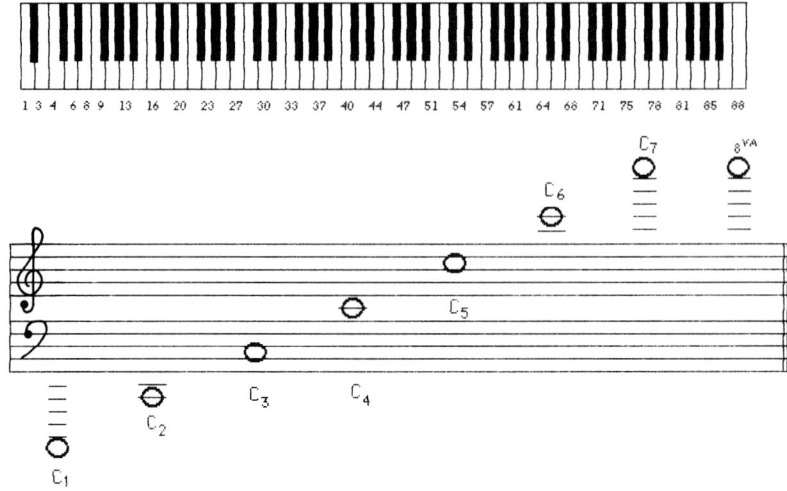

Figure B.1 The numbered keys of a piano together with the corresponding musical notation for all the 'C' notes.

> In all systems the designation applies not only to C but to all notes of the octave immediately above it. Thus in the American Standard , middle C is designated C_4 and the subscript 4 applies to all the notes immediately above it until B_4 is reached. Then the next note above that is C_5 and so on.

Table B.1

Piano note number	4	16	28	40	52	64	76	88
American Standard	C_1	C_2	C_3	C_4	C_5	C_6	C_7	C_8
American Alternative	C_3	C_2	C_1	C	C^1	C^2	C^3	C^4
Helmholtz	CC	C	c	c'	c''	c'''	c''''	c^v
English Organ	CCC	CC	C	c	c'	c''	c'''	c''''
ET frequency (Hz)	32.7	65.4	130.8	261.6	523.3	1046.5	2093	4186

B.3 RATIOS FOR THE JUST DIATONIC SCALE

In section 5.2 we saw that the scale of Just intonation involves three different intervals, the major tone, the minor tone and the semi tone and the sequence for the major diatonic scale was major, minor, semi, major, minor, major, semi, with ratios

9/8 10/9 16/15 9/8 10/9 9/8 16/15.

We can find the ratio of each note to the first by multiplying the successive ratios as follows

1/1 9/8 5/4 4/3 3/2 5/3 15/8 2/1.

So we can now write out the frequencies in Hertz for the Just scale on C that will include the note A_4 as 440 Hz.

264 297 330 352 396 440 495 528
C_4 D_4 E_4 F_4 G_4 A_4 B_4 C_5.

B.4 THE USE OF CENTS IN FREQUENCY MEASUREMENT

Since frequency is a logarithmic quantity (that is each successive *doubling* seems to the ear to be an equal *increment*) and through the use of the equal tempered scale, in which the octave is divided into twelve equal semitones, it becomes possible to specify small differences in frequency by means of a new unit called the 'cent'. We define a semitone as consisting of 100 cents and so the octave is 1200 cents. But notice that cents are not related to absolute frequency, but to intervals. Thus the interval of an equal tempered fifth is 700 cents and an equal tempered fourth is 500 cents. Just as intervals can be multiplied, cents being logarithmic can be added. So if we add a fifth (e.g., C_4 to G_4) to a fourth (e.g., G_4 to C_5) we get an octave (in this example C_4 to C_5) and, adding the cents, we obtain 700 + 500 = 1200.

Mathematically we can say that if the ratio of two notes is A/B then the interval in cents is:

$$1200\frac{\log_{10} A/B}{\log_{10} 2}.$$

This provides us with an interesting way of displaying the difference between the Just scale and the equal tempered scale as follows:

Table B.2

Note	Frequency (J)	Frequency (ET)	Interval from C_4 (J, in cents)	Interval from C_4 (ET, in cents)
C_4	264	261.6	0	0
D_4	297	293.6	204	200
E_4	330	329.6	386	400
F_4	352	349.2	498	500
G_4	396	392	702	700
A_4	440	440	884	900
B_4	495	493.9	1088	1100
C_5	528	523.2	1200	1200

Finally we give the complete list of frequencies of Just and equal tempered chromatic scales (that is including sharps and flats) starting on C. Notice that, in the Just scale, each note has both a sharp and a flat and the ratios are 24/25, 1, 25/24 (approximately 70 cents on either side of the note), whereas in the equal tempered scale the sharp of any given note is averaged with the flat of the note above.

Table B.3

Frequencies							
Just scale				Equal tempered scale			
C_4	264	$G^\flat$	380	C_4	261.6	B	495
C#	275	G	396	$C\#/D^\flat$	277.2	C_5	523.3
$D^\flat$	285	G#	412	D	293.6		
D	297	$A^\flat$	422	$D\#/E^\flat$	311.1		
D#	309	A	440	E	329.6		
$E^\flat$	317	A#	458	F	349.2		
E	330	$B^\flat$	475	$F\#/G^\flat$	370		
$F^\flat$	338	B	495	G	392		
E#	344	$C^\flat$	507	$G\#/A^\flat$	415.3		
F	352	B#	516	A	440		
F#	367	C_5	528	$A\#/B^\flat$	466.2		

Acknowledgments

In addition to the acknowledgments made in the introduction there are many people who helped to make both the lectures, and this publication possible and to whom a large debt of gratitude is owed.

Shell (UK) provided generous sponsorship for the televising of the five lectures on which the book is based and also provided still photographs which have been used as illustrations in the book. (Details of the figures involved are given elsewhere.)

InCA (Independent Communications Associates) were responsible for recording the programmes with William Woollard as Executive Producer, Brian Johnson as Producer and Richard Thomson as Associate Producer. I am particularly grateful to them for help in researching all kinds of details and for making the whole exercise as smooth as such a complicated operation could be.

Dr Bryson Gore, Mr Bippin Panjevane, Mrs Irena McCabe and many other colleagues from the Royal Institution gave their usual unstinting help and support both in the preparations for the lectures and in their execution.

Dr John Bowsher of the University of Surrey, Dr Richard Smith (trumpet maker), Dr Bernard Richardson of Cardiff, David Marshall of Roland (UK), Andrew Small of B & K, Jan Osman of Boosey & Hawkes, David Bristow and Michael Barnes of Yamaha (R & D), Professor David Pye of Queen Mary & Westfield College, Dr John Tyrer of the University of Technology at Loughborough, the Rector and the Director of Music at St. James Church, Spanish Place, and the BBC Radiophonic Workshop all gave valuable help in performing some of the quite complicated demonstrations that added so much interest to the lectures and which have been described in the book. Russell Johnson and Artec Consultants Inc., who were responsible for the design of Symphony Hall at Birmingham, kindly made details of this fascinating building available to me.

Many companies and organisations loaned equipment for use in demonstrations without which the lectures would have been very difficult to perform; much of the apparatus is mentioned and illustrated in the book. My warmest thanks go to: the Royal Institution of Great Britain, the BBC,

The Department of Physics of the University of Wales, Cardiff, The Musical Instrument Department of the London College of Furniture, Boosey and Hawkes (MI) Ltd, Brüel & Kjaer (UK) Ltd, Yamaha–Kemble Music (UK) Ltd, Yamaha R & D (London), Roland (UK) Ltd.

Robert Watkins, former photographer in the Department of Physics at Cardiff, has given enormous amounts of help and advice over the years for which I am immensely grateful; many of the original photographs in the book were produced by him.

The following people generously made available to me help and advice of all kinds and I owe them a great debt of gratitude:-

> Dr Michael Greenough, Dr George Harburn, Dr Bernard Richardson, and Dr Phillip Williams of the Department of Physics of the University of Wales at Cardiff, Dr Frances Palmer of the Horniman Museum, Dr Eric Clarke, Dr Simon Emmerson and Dr Malcolm Troup of the Music Department of the City University. I hope that I have included all, but the lectures were being contemplated over a period of more than a year and it is difficult to remember all the sources of ideas. To any who have been inadvertently omitted I offer my sincere apologies.

When the work on the manuscript was complete, my former colleague from the Department of Physics at the University of Wales College of Cardiff, Dr R Phillip Williams, generously gave up time to read the whole and his very critical eye detected a great many errors and infelicities which were corrected. I am most grateful to him. It goes without saying that any remaining errors are entirely my responsibility.

I am also grateful to Maureen Clarke, Al Troyano and their colleagues at Institute of Physics Publishing Ltd in Bristol who produced the final book with their usual helpful efficiency.

FIGURE ACKNOWLEDGMENTS

The following sources kindly granted permission to reproduce figures for this book.

Artec Consultants Inc – fig 6.16

Benade Arthur H 1976 *Fundamentals of Musical Acoustics* (New York: Oxford University Press) – fig 4.24

BBC Radiophonic Workshop – fig 5.10

Brüel & Kjaer (UK) Ltd – fig 2.23

Catgut Acoustical Society Inc – fig 3.22

Taylor C A 1965 *The Physics of Musical Sounds* (London: Edward Arnold) – figs 1.15, 1.19, 1.20, 1.21, 2.27, 4.28(*a*), (*b*) and (*c*)

Taylor Charles 1979 *Sound in the Sixth*, *Phys. Educ.* **14** – figs 2.12, 4.7
Taylor Charles 1988 *The Art and Science of Lecture Demonstration* (Bristol: Hilger) – fig 3.2
Taylor Charles 1990 *Music and the Acoustics of Buildings*, *Phys. Educ.* **25** – figs 6.7, 6.8
Dr Eric Clarke, The City University – fig 3.29
Russell Johnson – fig 6.17
Kinsler L E and Frey A R 1962 *Fundamentals of Acoustics* (New York: Wiley) – figs 1.11, 1.12, 1.22
Meyer Jürgen 1978 *Acoustics and the Performance of Music* (Frankfurt: Verlag Das Musikinstrument) – fig 4.15
Premier Percussion – fig 2.8
Dr Bernard Richardson, University of Wales, College of Cardiff – figs 3.5, 3.6, 3.12
The Royal Albert Hall – fig 6.11
From the Proceedings of The Royal Institution of Great Britain: vol 48 – fig 3.9(*a*); vol 52 – figs 2.9, 3.1, 3.14, 3.15, 3.18, 3.19; vol 61 – figs 2.2, 2.24, 3.8
The Royal Society, *Science and Public Affairs* vol 3 – figs 2.11, 2.31, 3.3, 3.11, 4.16
Shell Photographic Service – frontispiece, figs 2.3(*b*), 2.10, 3.16, 3.17, 3.20, 3.21, 4.4, 4.23
Dr John Tyrer, University of Technology, Loughborough – fig 3.13
Yamaha–Kemble Music (UK) Ltd – fig 5.9

The following were taken from the fourth edition of John Tyndall's book *Sound*, published in 1883: figs 1.1, 1.2, 2.7.

The following were taken from my book *Sounds of Music* which was the record of my 1971 Cristmas Lectures: figs 1.9, 1.10, 1.13, 1.16, 2.16, 2.18, 2.22, 2.24, 2.26, 2.28, 2.30, 2.33, 2.34, 3.23, 3.24, 3.25, 4.10, 4.12, 4.13, 4.14, 4.17, 4.18, 4.19, 4.20, 4.21, 4.26, 4.27, 4.30, 4.31, 5.1, 5.2, 5.4, 5.5, 5.6, 6.1, 6.2, 6.3, 6.4, 6.5, 6.6, 6.9, 6.10, 6.11, 6.12, 6.13, 6.14, 6.15, A.1.

Bibliography and Suggestions for Further Reading

Backus John 1977 *The Acoustical Foundations of Music* (New York: W W Norton)

Benade Arthur H 1976 *Fundamentals of Musical Acoustics* (New York: Oxford University Press)

Benade Arthur H *Horns, Strings and Harmony* (New York: Doubleday)

Berg R E and Stork D G 1982 *The Physics of Sound* (New Jersey: Prentice-Hall)

Cremer L 1984 (translated by J S Allen) *The Physics of the Violin* (Cambridge, MA: MIT Press)

Donington Robert 1982 *Music and its Instruments* (London: Methuen)

Gill Dominic 1984 *The Book of the Violin* (Oxford: Phaidon)

Helmholtz H L F 1954 *On the Sensations of Tone* (New York: Dover) reprint of 1875 edition

Hutchins C M 1975 & 1976 (ed) *Musical Acoustics Parts 1 & 2 (Violin)*, Benchmark Papers in Acoustics vols 5 & 6 (Dowden, PA: Hutchinson & Ross, Inc.)

Johnston Ian 1989 *Measured Tones* (Bristol: Hilger)

Kennedy Michael 1985 *The Oxford Dictionary of Music* (Oxford: Oxford University Press)

Kent Earle L 1977 (ed) *Musical Acoustics (Piano & Wind)*, Benchmark Papers in Acoustics vol 9 (Dowden, PA: Hutchinson & Ross, Inc.)

Meyer Jürgen 1978 *Acoustics and the Performance of Music* (Frankfurt: Verlag Das Musikinstrument) English edition

Miller D C 1935 *Anecdotal History of the Science of Sound* (New York: Macmillan)

Moore B J C 1977 *Introduction to the Psychology of Hearing* (Cambridge: Cambridge University Press)

Northwood Thomas D 1977 (ed) *Architectural Acoustics*, Benchmark Papers in Acoustics vol 10 (Dowden, PA: Hutchinson & Ross, Inc.)

Parsons D 1975 *Directory of Tunes and Musical Themes* (Cambridge: Spencer Brown)

Pierce J R 1983 *The Science of Musical Sounds* (New York: Scientific American Books)

Richardson Bernard E 1988 *Vibrations of Stringed Musical Instruments (University of Wales Review)* Autumn

Roederer J G 1979 *Introduction to the Physics & Psychophysics of Music* (New York: Springer)

Taylor C A 1965 *The Physics of Musical Sounds* (London: English Universities Press)

Taylor Charles 1975 *John Tyndall's Demonstrations on Sound, 1854–1882, Proc. R. Inst. Great Britain* **48**

Taylor Charles 1976 *Sounds of Music* (London: BBC)

Taylor Charles 1978 *Sounds of Music*, *Sci. Prog. Oxf.* **65**

Taylor Charles 1979 *Sound in the Sixth*, *Phys. Educ.* **14**

Taylor Charles 1980 *Science and the Violin Family*, *Proc. R. Inst. Great Britain* **52**

Taylor Charles 1988 *The Art and Science of Lecture Demonstration* (Bristol: Hilger)

Taylor Charles 1986 *A Scientist in the World of Music*, *Sci. Publ. Affairs* **3**

Taylor Charles 1989 *Physics and Music Twenty Years On*, *Proc. R. Inst. Great Britain* **61**

Taylor Charles 1990 *Music and the Acoustics of Buildings*, *Phys. Educ.* **25**

Tyndal John 1883 *Sound* 4th edn (London: Longmans Green)

Index

ADC, analogue–digital converter, 199
Amati, 90, 103
Analogue synthesis, 186–95
Analysis, frequency, of sounds, 65–7
Anechoic chamber, 225, 228
Artec Consultants Inc., 232
Audibility, limits of, 10

BBC Maida Vale studios, 227–30
BBC Radiophonic Workshop, 188, 194–5, 204–6
Bach, J S, 183, 185
Bagpipe chanter, 64–5, 145
Bagpipe reeds, 145
Baschet, F, 79
Bass bar, 104
Bassoon, 37, 141–2
Beats, 31–7
Bell Telephone Laboratories, 199
Benade, A H, 71, 138, 150–4, 162
Berlioz, H, 29
Birmingham's Symphony Hall, 232–4
Blocks, wooden, music from, 15, 45, 50
Body of stringed instrument, 99–103
Boehm, T, systems for woodwind, 147, 154–6
Bowden, Lord, 227
Bowing, 75–9, 89, 103–5
Bowsher, J, 138, 165
Boyle, Robert, 4
Brain, function of, in hearing, 8–10, 16–27, 34–41, 81–6, 225–7
Brass family, 62, 160–5
Bridge, Tacoma, 60
Brüel & Kjaer Ltd., 67

Catgut Acoustical Society, 115–9
Cent, as a measure of frequency, 243–4
Chalumeau register of clarinet, 71, 140
Chladni plate, 91–5, 109, 239
Chorus effect, 226
Chowning, John, 201
Clarinet, 71, 81–3, 140–1, 148, 152–3
Clarino register of clarinet, 71, 140
Clarke, Eric, 127–8
Clavichord, 124–5
Clock, pendulum, 47–8
Compton Electrone organ, 186–9
Computer, as musical instrument, 187, 199–200
Concert Hall, testing musical instruments in the, 27, 90, 111–4, 128
Concert Halls, acoustics of, 211 *et seq*
Conical pipes, 138–40
Cooperation between modes in woodwinds, 152–3
Cor Anglais, 146
Cornemuse, 63–5, 145
Cornett, 158–60
Corrugated tube, notes produced by whirling a, 134–6
Critical bands in hearing, 19, 142–3
Crooks in brass instruments, 162–4
Crumhorn, 145
Crwth, 104–5

DAC, digital–analogue converter, 199
Decibels, 19–24
Diatonic scale, 181–5, 242–3
Didgeridoo, 56
Difference tone, 33–7, 141–3
Digital FM synthesis, 201
Digital techniques, 197–205
Discord, origins of, 30–4
Disklavier, 127, 204
Dissonance curve, Helmholtz, 30, 35
Domes, acoustic problems of, and their corrective treatment, 221–5

Drums, talking, 27–8

Ear
  frequency range of, 10–12
  mechanism of, 10–12, 16–19, 30–9
  non–linearity in the, 34–5, 141
Ear–brain system, 9, 16, 27, 34, 39, 80–3, 226
Echoes in buildings, 221–4
Eddies, 58–61
Edge tones, 58–61, 131, 133–5, 143–5, 157–8, 167
Electrone, Compton, 81, 186–9
Electronic organs, 81, 186–9, 205–6
Envelope of a note, 83–5, 188–95
Equal temperament, 182–6, 244
Escapement, 47–8
Euphonium, 165
Eustachian tube, 18

F-holes, 104, 116
Families of musical instruments
  first, 15–16, 47, 165, 169
  second, 52
  third, 61
Fant, G, 176
Faraday, Michael, 48, 93
Fechner, Weber–Fechner Law, 19–20
Fifth, as a natural musical interval, 180–2
Flame, sensitive, 12–13
Flame, singing, 48–50
Fletcher–Munson lines of equal subjective loudness, 23
Flute, 82, 144–5
Flute, key system of a, 154–6
Flutter effect, 221
FM, frequency modulation in synthesis, 201
Fogg lecture theatre at Harvard, 216–8
Fork, tuning, 52–4, 98
Formant characteristic, 55, 111, 125
Formants in speech, 173–8
Fourier's theorem, 72, 81, 121
Free Trade Hall, Manchester, 230
Frequency analyser, real time, 67–8
Frequency and pitch, 10–15, 39–41, 241–4
Frequency range of the ear, 10–12
Frequency, natural 45–6
Frequency, privileged, 136–40, 165
Frets, 100, 103

Fundamental, missing, 37, 141

Galilei, Galileo, 46
Glasses, musical, 76–9
Grain, effect of wood, 105–9
Guarneri, 90, 103
Guimbard, or Jew's harp, 175
Guitar, 81–2, 99–103, 119–21

Hammond organ, 81, 186–9
Harmonic, meaning of term, 35, 61–3
Harmonics in pipes, 133–43, 152–3, 156–8
Harmonics of vibrating strings, 72–5, 119–23
Harmonics, aural, 36
Harmonics, production by reeds, 67–8
Harmony and discord, 30–4
Harpsichord, 84, 121–4
Hauksbee, Francis, 4
Hearing, measurements on, 19–24
Helmholtz, H L F, 15, 37, 65–6, 98–9, 104, 161, 179, 241
Hertz, unit of frequency measurement, 11
Higgins, Dr, 48
High register of clarinet, 71, 140
Hoffnung, Gerard, 43
Hologram method of testing instruments, 92–5, 99–103, 237–9
Horn, baritone, 165
Hutchins, Carleen M, 115–9

Impedance, input, of pipes, 156–8
Information in music, 27–8
Inharmonic partials of piano strings, 126
Initiation by blowing, 58–65
Initiation by bowing, 75–9, 103–5
Initiation by plucking, 119–24
Interference, 31, 237–9
Intervals, equal tempered, 182–5, 243–4
Intervals, Just, 180–5, 242–4

Jew's harp, 175
Johnson, Russell, 232–4
Just intonation
  intervals in, 180–5, 242–4
  origin of scale of, 179–82

Key, in musical compositions, 179–85
Keyboard instruments, 121–8, 204
Keys on woodwind instruments, 147, 154–6
König, Rodolf, 15
Kreisler, Fritz, 111

Laser holograms for testing instruments, 92–5, 99–103, 237–9
Linear and non–linear behaviour, 35, 141, 151
Loudness curves for a violin, 109–10
Loudness versus intelligibility in building acoustics, 214–6
Loudness, curves of equal subjective, 23

Major diatonic scale, 181–5
Masking, 142–3
Matthews, Max, 110, 199
Mel, unit of melodic pitch, 41
Meyer, Jürgen, 148, 226
MIDI, 187, 203–9
Minor diatonic scale, 184–5
Modes, cooperation between, in reed instruments, 152–3
Modes of vibration
  of a stretched string, 72–5
  of hollow bodies, 95–9
  of plates, 91–5, 109, 239
Morse code, 27
Music V computer programme, 199
Musique Concrète, 187, 195, 200

Nodes on musical glasses, 76
Nodes on plates, 91–5, 109, 239
Noise in buildings, 234–6
Non-harmonic partials of piano strings, 126
Non-linear behaviour, 35, 141, 151
Notation, systems of pitch, 11–12, 241–2
Note, starting a, 43, 80–3
Notes changing with time, 80–6

Oboe, 145–7, 154
Octave as natural musical interval, 11, 180
Ondes Martenots, 186
Open window unit in acoustic absorption measurement, 217–8
Ophicleide, 159
Organ at St. James', Spanish Place, London, 171
Organ
  Compton Electrone, 186–9
  Hammond, 186–9
  mechanism of an, 170–3
Organ pipe, resultant bass, 34, 170
Organ pipes, 12, 165–70
Organs, temperament of, 185
Oscillator, electronic as a sound source, 186–95
Overtone, meaning of the term, 62

Padgham, Charles, 185
Paget, R, 175
Pain, threshold of, 19
Panpipes, 58–9
Parsons, Denys, Directory of Tunes, 28
Partial, meaning of the term, 62
Pendulum models, 47, 67–70
Perception of harmonic mixtures, 68–72
Period of an oscillator, 14
Phon, unit of equivalent loudness, 22–4
Piano touch, 127–8
Pianoforte, 84, 124–8, 204
Pibcorn, 145
Pierce, John, 199
Pipes
  organ, 12, 165–70
  vibrations in conical, 138–40, 156–8
  vibrations in cylindrical, 58–61, 132–8, 156–8
Pitch and frequency, 10–15, 39–41, 241–4
Pitch
  conventions for specifying, 11, 168, 241–2
  dynamic relationship with amplitude, 39–41
Plucking, influence on tone of point of, 72–5, 119–21
Precedence effect, 226
Pressure, magnitude of variation in sound wave, 19–24
Privileged frequencies, 136–40, 165
Proprioceptive response, 111
Psychological aspects of hearing, 8–10, 16–27, 34–41, 81–6, 225–7

Pure tones, 13–15
Purfling on a violin or 'cello, 108
Pye, David, 13

Radiation from instruments, 111–6, 147–9
Raman, Sir C V, 78
Recipe for vibrations in wind instruments, 71, 140–3, 149–54, 165
Recorders, 143–5, 152–4
Reeds, 63–5, 67–8, 140–3, 145–7, 151, 160–2, 169–70
Residue, tonal, 34–7, 141–3
Resonance, 48–56, 60, 65–7, 97–104
Reverberation
  artificial, 230–2
  time of, 216–20, 230–2
Richardson, Bernard, 94, 100, 107, 112–4
Rijke's tube, 58
Ripple tank used in acoustic design, 227
Rods, brass, longitudinal vibrations in, 77–8
Roland sampler, 196
Roland synthesiser, 202
Royal Albert Hall, modification of acoustics in the, 222–3
Royal Festival Hall, modification of acoustics in the, 232
Royal Institution, the, 8, 65, 66, 74, 190

Sabine, W C, 216–8
Sampling, 195–7
Sansa, West-African instrument, 16
Saunders, F A, 110–1, 115
Saw, musical, 94–6
Scale of Just intonation, 179–82, 242–4
Scale, equal tempered, 182–6, 244
Scales
  major and minor, 181–5, 242–4
  musical, 179–86, 242–4
Schaeffer, Pierre, 195
Schouten, J F, 143
Sensitive flame, 12–13
Sequencer, 187, 190–5
Serpent, 158–60
Shantu, West-African instrument, 44–5
Side-holes, function in woodwinds, 143–56
Sine wave, 13–14
Singing flame, 48–50
Smith, Dave, originator of MIDI, 203
Sone, unit of subjective loudness, 24–5
Sound level, 19–24
Sound post in a violin or 'cello, 104–6
Sounds of Music, 1971 Christmas lectures, 1, 190, 195, 212, 241
Speech, formants in, 173–8
Speech, x-rays of vocal tract during, 176
Spinet, 121–4
Starting transients, 55, 83–7
Stick-slip motion, 76–9
Stradivari, Antonio, 90, 103, 109
Straws, panpipes and reeds made from, 58–61, 63–5
Stringed instruments, 53–4, 89 *et seq*
Strings, vibrations of, 72–9, 119–24
Structures Sonores, 79–80
Sydney Opera House organ, 173
Symphony Hall, Birmingham, 232–4
Synthesis, analogue, 186–95
Synthesis, digital, 197–205
Syrinx, 58–9

Tabor and tabor pipe, 63, 158
Tacoma Bridge disaster, 60
Tape recorder, 86, 187, 195, 200, 216
Telharmonium, 186
Temperament, 182–6, 244
Theremin, 186
Time of reverberation, 216–20, 230–2
Tonal complex, 36, 143
Tonal residue, 34–7, 141–3
Tone, pure, 13–15
Touch, piano, 127–8
Transient, starting, 55, 83–7
Trombone, 138, 162–5
Trumpet, 160–5
Tube, behaviour of compression at end of, 132–3
Tube, eustachian, 18
Tubes
  notes produced by whirling corrugated, 134–6
  vibrations in conical, 138–40
  vibrations in cylindrical, 58–61, 132–8, 156–8
Tuning fork, 52, 54, 98

Tyndall, John, 4, 6–8, 12, 48–52, 58, 77, 97
Tyrer, John, 102

Ultrasonics, 12–13, 15
University of Manchester Institute of Science & Technology, 231

Vacuum, demonstration that sound will not travel through a, 5
Valves on brass instruments, 162–5
Varnish on violins and 'cellos, 109
Vibration recipes, 71, 140–3, 149–54, 165
Vibrato, 31–2, 51, 81–3, 170
Viol, 89–90
Violin, 89–91, 103–19
Virginals, 121–4
Voice, 173–8
Voltage control in synthesisers, 187–95

Wavelength, 13–15
Weber–Fechner Law, 19–20
Wheatstone, Sir Charles, 8, 76
Wind-cap instruments, 63–5, 145
Wolf tone, 114–5

Xylophone, 15, 51–2

Yamaha Disklavier, 127, 204
Yamaha synthesisers, 200–3
Yamaha wind controller, 204
Young, Thomas, 227

Zinoviev, Peter, 190